Krav Maga

Defenses Against Various Attacks

Logbook

Shawn Kovacich

AUTHOR APPROVED SHOOT/NO SHOOT EDITION

Printed in the United States of America

Library of Congress Control Number: 2019912090

ISBN #978-1-7336809-6-7

Table of Contents

Introduction

Disclaimer....vii

Safety....viii

And Finally....viii

How To Use This Logbook....ix

Cover and Spine....ix

Logbook Entry....ix

Defense Against....ix

Start Date....ix

Notes....ix

Slow Reps....x

Reg Reps....x

Notes....x

End Date....x

Code Key....x

Code Key....xii

Logbook Entry (write down the name of the attack you are defending against)

____________________ 1

____________________ 3

____________________ 5

____________________ 7

____________________ 9

____________________ 11

____________________ 13

____________________ 15

____________________ 17

____________________ 19

____________________ 21

____________________ 23

____________________ 25

____________________ 27

____________________ 29

____________________ 31

____________________ 33

____________________ 35

Logbook Entry (continued)

______ 37
______ 39
______ 41
______ 43
______ 45
______ 47
______ 49
______ 51
______ 53
______ 55
______ 57
______ 59
______ 61
______ 63
______ 65
______ 67
______ 69
______ 71
______ 73
______ 75
______ 77
______ 79
______ 81
______ 83
______ 85
______ 87
______ 89
______ 91
______ 93
______ 95
______ 97
______ 99

Additional Notes....................................101

Disclaimer

Please note that the author, Shawn Kovacich, and the publisher, **Shoot-NO-Shoot Productions**, Chikara Kan, Inc., of this logbook are **Not Responsible** in any manner whatsoever, for any injury, which may occur by reading and/or following the instructions located within this logbook. The distributor and any and all other individuals and or businesses connected with the production and/or distribution of this book also disclaim any and all liability for any damage or injury of any kind that may result from the appropriate use or inappropriate misuse of any information contained within this book.

The techniques that you will be executing while utilizing this logbook are potentially lethal in nature and are intended to be used in a **safe and law abiding manner** during practice and also if/when confronted with a potentially life threatening situation that may or may not require the use of lethal force in order to protect ones life. It is the reader or readers direct responsibility to research and comply with all applicable laws regarding the use of force in a self-defense situation in the areas in which they reside, do business, or frequent. **Ignorance of the law never has been, nor is it likely that it ever will be, a viable defense.**

Due to the fact that the techniques described in this book are physical in nature, it is essential that the reader or readers of this book consult a qualified and competent physician before following any of the activities, physical or otherwise, which are described within this book.

This book and all the information contained therein is intended to be used as a supplemental training aid, and should be used as such, under the guidance of a qualified and competent martial arts/self-defense instructor.

This book and the information contained within are for informational purposes only!

Copyrights & Trademarks

SAFETY

Due to the physical nature of the activities that you will be participating in while utilizing this logbook, it is highly recommended that you first consult a qualified and competent physician before starting any physical training routine. Regardless of your current physical condition. Remember, an ounce of prevention is worth a pound of cure. And also remember, **Safety First, Foremost, and Always!**

AND FINALLY...

You would do well to remember that progress should be measured in inches, not in miles. You are not going to execute 10 or 20 defenses against a particular attack and then all of a sudden be showing a remarkable improvement in your self-defensive skills. It takes time, proper practice, full immersion and active participation, and consistency in practice in order to achieve positive results. You probably won't see any big changes after just a few dozen repetitions, but you should be seeing some significant improvement after the first 100 repetitions defending against a particular attack. And, you should continue to show improvement after every few hundred repetitions that you execute, if you have fully immersed yourself both physically and mentally, while actively participating in each and every training session.

In addition to that, if you have also been studying and refining your defenses against various attacks under the watchful eye of a qualified and competent martial arts/self-defense instructor, then you will make even greater progress in honing your self-defensive skills.

HOW TO USE THIS LOGBOOK

In order to obtain the full benefits of this **Krav Maga: Defenses Against Various Attacks Logbook**, you will want to follow these simple instructions on how to best utilize this logbook. This logbook should be used along with your active participation with the various defenses that you are currently practicing. These defenses that you have learned and/or are learning, should be practiced on a regular and consistent basis while keeping an extensive and detailed record of the number of repetitions that your are executing against each particular attack.

However, please make sure that you are first doing the defenses correctly before you start keeping track of the thousands of repetitions that you are going to start keeping track of. Because if you are doing the defenses incorrectly the first time, and you continue to practice them incorrectly, how do you think you are going to execute them the 6,000th time?

COVER & SPINE

When you look at the front cover and the spine, you will see a small white box in the lower left hand corner on the front cover and a smaller white box on the spine. These are include in order to give you a space in which to write down the volume number or other such indicator of the logbook that you are currently using.

LOGBOOK ENTRY (on the **"Table of Contents"** pages)

This is where you will write down the name of the **Attack** that you are **"Defending Against."**

DEFENSE AGAINST

Write down the name of the **Attack** that you are **"Defending Against"** and will be practicing and keeping meticulous track of. For example; Straight Punch, Tackle, Baseball Bat, Rifle, etc.

START DATE

This is where you will enter the **"Date"** that you started keeping track of the number of repetitions that you executed **"Defending Against"** this particular attack. Be sure and include the month, day, and year.

NOTES

This section allows you to write down any particular **"Notes"** that you may have concerning this **"Defense Against"** a particular attack that you are keeping track of.

SLOW REPS

This section allows you to keep track of the number of slow repetitions or **"Slow Reps"** that you execute throughout the time frame that you are also executing your regular repetitions of this **"Defense Against"** this particular attack. My personal recommendation is that you should execute at least 3 to 5 **"Slow Reps"** at the beginning and at the end of every session that you are working on this defense against this particular attack.

For example if you are going to execute 50 defenses against a baseball bat. You should start out be executing 3 to 5 slow repetitions followed by 50 repetitions at normal speed and then finish up with another 3 to 5 slow repetitions. That would give you a total of 56 to 60 defenses executed in that particular training session, 6 to 10 **"Slow Reps"** and 50 **"Reg Reps"** for that **"Defense Against"** this particular attack.

REG REPS

The regular repetitions or **"Reg Reps"** section allows you to circle or check off the number of regular repetitions of this particular **"Defense Against"** this particular attack that you are keeping track of in increments of 10. So, for every 10 repetitions that you execute of a **"Defense Against"** a particular attack, you would check off the next 10 repetition increment. For example; if you are just starting out and you just finished executing say 50 repetitions of a **"Defense Against"** a baseball bat attack, you would then circle or check off the following numbers; 10, 20, 30, 40, and 50.

NOTES

This **"Notes"** section, allows you the opportunity to write down any particular notes that you may have concerning this **"Defense Against"** a particular attack after every 1,000 repetitions executed. There is no notes section after the **"Slow Reps,"** as your slow repetitions are intended to be executed at various stages during the execution of the 5,000 regular repetitions or "**Reg Reps.**" Be sure and include the month, day, and year. Use the **"Additional Notes"** section included at the end of this logbook to keep track of any additional items that you feel you need to keep track of during the course of utilizing this logbook.

END DATE

This is where you will enter the **"Date"** in which you completed the 1,000 slow repetitions **"Slow Reps"** and the 5,000 regular repetitions **"Reg Reps"** of this **"Defense Against"** a particular attack.

CODE KEY

This section was included in order to allow you the opportunity to personalize the way you keep track of your progress by allowing you to use a variety of symbols, ink colors, etc., in order to keep track of any particularities that you may want to include in your own personal way to add to your training regiment. I have included some various possibilities that you could use, or come up with your own that may fit your own particular needs. So without further ado, let's get started.

BLACK INK = Physically and mentally an average training routine.

BLUE INK = Physically and mentally not feeling well. Not a typical training routine.

RED INK = Really pumped up both physically and mentally. Outstanding training routine.

X (over the number) = Executed that particular number of repetitions flawlessly with no mistakes, etc.

CIRCLE (around the number) = Executed that particular number of repetitions with a few minor mistakes.

SQUARE (around the number) = Executed that particular number of repetitions with a good defense, but my counters were off.

TRIANGLE (around the number) = Executed that particular number of repetitions with a weak defense and weak counter-attack.

CODE KEY

Here is the **"Code Key"** for you to personalize to your own particular needs and training regiment. I would strongly recommend that you read through the **"How To Use This Logbook"** section prior to using this logbook in order to maximize its use to its fullest potential. Be sure and fill this out prior to starting your logbook.

BLACK INK ______________________________

____

BLUE INK ______________________________

____

RED INK ______________________________

____

X ______________________________

____

CIRCLE ______________________________

____

SQUARE ______________________________

____

TRIANGLE ______________________________

____

Defense Against: ______________________________ **Start Date:** __________

Notes: ____

____

____

____

Slow Reps: 10 - 20 - 30 - 40 - 50 - 60 - 70 - 80 - 90 - 100 - 110 - 120 - 130
140 - 150 - 160 - 170 - 180 - 190 - 200 - 210 - 220 - 230 - 240 - 250 - 260
270 - 280 - 290 - 300 - 310 - 320 - 330 - 340 - 350 - 360 - 370 - 380 - 390
400 - 410 - 420 - 430 - 440 - 450 - 460 - 470 - 480 - 490 - 500 - 510 - 520
530 - 540 - 550 - 560 - 570 - 580 - 590 - 600 - 610 - 620 - 630 - 640 - 650
660 - 670 - 680 - 690 - 700 - 710 - 720 - 730 - 740 - 750 - 760 - 770 - 780
790 - 800 - 810 - 820 - 830 - 840 - 850 - 860 - 870 - 880 - 890 - 900 - 910
920 - 930 - 940 - 950 - 960 - 970 - 980 - 990 - 1000

Reg Reps: 10 - 20 - 30 - 40 - 50 - 60 - 70 - 80 - 90 - 100 - 110 - 120 - 130
140 - 150 - 160 - 170 - 180 - 190 - 200 - 210 - 220 - 230 - 240 - 250 - 260
270 - 280 - 290 - 300 - 310 - 320 - 330 - 340 - 350 - 360 - 370 - 380 - 390
400 - 410 - 420 - 430 - 440 - 450 - 460 - 470 - 480 - 490 - 500 - 510 - 520
530 - 540 - 550 - 560 - 570 - 580 - 590 - 600 - 610 - 620 - 630 - 640 - 650
660 - 670 - 680 - 690 - 700 - 710 - 720 - 730 - 740 - 750 - 760 - 770 - 780
790 - 800 - 810 - 820 - 830 - 840 - 850 - 860 - 870 - 880 - 890 - 900 - 910
920 - 930 - 940 - 950 - 960 - 970 - 980 - 990 - 1000

Notes: ____

____

____

____

Reg Reps: 10 - 20 - 30 - 40 - 50 - 60 - 70 - 80 - 90 - 100 - 110 - 120 - 130
140 - 150 - 160 - 170 - 180 - 190 - 200 - 210 - 220 - 230 - 240 - 250 - 260
270 - 280 - 290 - 300 - 310 - 320 - 330 - 340 - 350 - 360 - 370 - 380 - 390
400 - 410 - 420 - 430 - 440 - 450 - 460 - 470 - 480 - 490 - 500 - 510 - 520
530 - 540 - 550 - 560 - 570 - 580 - 590 - 600 - 610 - 620 - 630 - 640 - 650
660 - 670 - 680 - 690 - 700 - 710 - 720 - 730 - 740 - 750 - 760 - 770 - 780
790 - 800 - 810 - 820 - 830 - 840 - 850 - 860 - 870 - 880 - 890 - 900 - 910
920 - 930 - 940 - 950 - 960 - 970 - 980 - 990 - 1000

Notes: ____

____

____

____

Reg Reps: 10 - 20 - 30 - 40 - 50 - 60 - 70 - 80 - 90 - 100 - 110 - 120 - 130
140 - 150 - 160 - 170 - 180 - 190 - 200 - 210 - 220 - 230 - 240 - 250 - 260
270 - 280 - 290 - 300 - 310 - 320 - 330 - 340 - 350 - 360 - 370 - 380 - 390
400 - 410 - 420 - 430 - 440 - 450 - 460 - 470 - 480 - 490 - 500 - 510 - 520
530 - 540 - 550 - 560 - 570 - 580 - 590 - 600 - 610 - 620 - 630 - 640 - 650
660 - 670 - 680 - 690 - 700 - 710 - 720 - 730 - 740 - 750 - 760 - 770 - 780
790 - 800 - 810 - 820 - 830 - 840 - 850 - 860 - 870 - 880 - 890 - 900 - 910
920 - 930 - 940 - 950 - 960 - 970 - 980 - 990 - 1000

Notes: ____________________

Reg Reps: 10 - 20 - 30 - 40 - 50 - 60 - 70 - 80 - 90 - 100 - 110 - 120 - 130
140 - 150 - 160 - 170 - 180 - 190 - 200 - 210 - 220 - 230 - 240 - 250 - 260
270 - 280 - 290 - 300 - 310 - 320 - 330 - 340 - 350 - 360 - 370 - 380 - 390
400 - 410 - 420 - 430 - 440 - 450 - 460 - 470 - 480 - 490 - 500 - 510 - 520
530 - 540 - 550 - 560 - 570 - 580 - 590 - 600 - 610 - 620 - 630 - 640 - 650
660 - 670 - 680 - 690 - 700 - 710 - 720 - 730 - 740 - 750 - 760 - 770 - 780
790 - 800 - 810 - 820 - 830 - 840 - 850 - 860 - 870 - 880 - 890 - 900 - 910
920 - 930 - 940 - 950 - 960 - 970 - 980 - 990 - 1000

Notes: ____________________

Reg Reps: 10 - 20 - 30 - 40 - 50 - 60 - 70 - 80 - 90 - 100 - 110 - 120 - 130
140 - 150 - 160 - 170 - 180 - 190 - 200 - 210 - 220 - 230 - 240 - 250 - 260
270 - 280 - 290 - 300 - 310 - 320 - 330 - 340 - 350 - 360 - 370 - 380 - 390
400 - 410 - 420 - 430 - 440 - 450 - 460 - 470 - 480 - 490 - 500 - 510 - 520
530 - 540 - 550 - 560 - 570 - 580 - 590 - 600 - 610 - 620 - 630 - 640 - 650
660 - 670 - 680 - 690 - 700 - 710 - 720 - 730 - 740 - 750 - 760 - 770 - 780
790 - 800 - 810 - 820 - 830 - 840 - 850 - 860 - 870 - 880 - 890 - 900 - 910
920 - 930 - 940 - 950 - 960 - 970 - 980 - 990 - 1000

Notes: ____________________

End Date: ____________

Defense Against:______________________ **Start Date:** __________

Notes:____

____

____

____

Slow Reps: 10 - 20 - 30 - 40 - 50 - 60 - 70 - 80 - 90 - 100 - 110 - 120 - 130
140 - 150 - 160 - 170 - 180 - 190 - 200 - 210 - 220 - 230 - 240 - 250 - 260
270 - 280 - 290 - 300 - 310 - 320 - 330 - 340 - 350 - 360 - 370 - 380 - 390
400 - 410 - 420 - 430 - 440 - 450 - 460 - 470 - 480 - 490 - 500 - 510 - 520
530 - 540 - 550 - 560 - 570 - 580 - 590 - 600 - 610 - 620 - 630 - 640 - 650
660 - 670 - 680 - 690 - 700 - 710 - 720 - 730 - 740 - 750 - 760 - 770 - 780
790 - 800 - 810 - 820 - 830 - 840 - 850 - 860 - 870 - 880 - 890 - 900 - 910
920 - 930 - 940 - 950 - 960 - 970 - 980 - 990 - 1000

Reg Reps: 10 - 20 - 30 - 40 - 50 - 60 - 70 - 80 - 90 - 100 - 110 - 120 - 130
140 - 150 - 160 - 170 - 180 - 190 - 200 - 210 - 220 - 230 - 240 - 250 - 260
270 - 280 - 290 - 300 - 310 - 320 - 330 - 340 - 350 - 360 - 370 - 380 - 390
400 - 410 - 420 - 430 - 440 - 450 - 460 - 470 - 480 - 490 - 500 - 510 - 520
530 - 540 - 550 - 560 - 570 - 580 - 590 - 600 - 610 - 620 - 630 - 640 - 650
660 - 670 - 680 - 690 - 700 - 710 - 720 - 730 - 740 - 750 - 760 - 770 - 780
790 - 800 - 810 - 820 - 830 - 840 - 850 - 860 - 870 - 880 - 890 - 900 - 910
920 - 930 - 940 - 950 - 960 - 970 - 980 - 990 - 1000

Notes: ____

____

____

____

Reg Reps: 10 - 20 - 30 - 40 - 50 - 60 - 70 - 80 - 90 - 100 - 110 - 120 - 130
140 - 150 - 160 - 170 - 180 - 190 - 200 - 210 - 220 - 230 - 240 - 250 - 260
270 - 280 - 290 - 300 - 310 - 320 - 330 - 340 - 350 - 360 - 370 - 380 - 390
400 - 410 - 420 - 430 - 440 - 450 - 460 - 470 - 480 - 490 - 500 - 510 - 520
530 - 540 - 550 - 560 - 570 - 580 - 590 - 600 - 610 - 620 - 630 - 640 - 650
660 - 670 - 680 - 690 - 700 - 710 - 720 - 730 - 740 - 750 - 760 - 770 - 780
790 - 800 - 810 - 820 - 830 - 840 - 850 - 860 - 870 - 880 - 890 - 900 - 910
920 - 930 - 940 - 950 - 960 - 970 - 980 - 990 - 1000

Notes: ____

____

____

____

Reg Reps: 10 - 20 - 30 - 40 - 50 - 60 - 70 - 80 - 90 - 100 - 110 - 120 - 130
140 - 150 - 160 - 170 - 180 - 190 - 200 - 210 - 220 - 230 - 240 - 250 - 260
270 - 280 - 290 - 300 - 310 - 320 - 330 - 340 - 350 - 360 - 370 - 380 - 390
400 - 410 - 420 - 430 - 440 - 450 - 460 - 470 - 480 - 490 - 500 - 510 - 520
530 - 540 - 550 - 560 - 570 - 580 - 590 - 600 - 610 - 620 - 630 - 640 - 650
660 - 670 - 680 - 690 - 700 - 710 - 720 - 730 - 740 - 750 - 760 - 770 - 780
790 - 800 - 810 - 820 - 830 - 840 - 850 - 860 - 870 - 880 - 890 - 900 - 910
920 - 930 - 940 - 950 - 960 - 970 - 980 - 990 - 1000

Notes: ____________________

Reg Reps: 10 - 20 - 30 - 40 - 50 - 60 - 70 - 80 - 90 - 100 - 110 - 120 - 130
140 - 150 - 160 - 170 - 180 - 190 - 200 - 210 - 220 - 230 - 240 - 250 - 260
270 - 280 - 290 - 300 - 310 - 320 - 330 - 340 - 350 - 360 - 370 - 380 - 390
400 - 410 - 420 - 430 - 440 - 450 - 460 - 470 - 480 - 490 - 500 - 510 - 520
530 - 540 - 550 - 560 - 570 - 580 - 590 - 600 - 610 - 620 - 630 - 640 - 650
660 - 670 - 680 - 690 - 700 - 710 - 720 - 730 - 740 - 750 - 760 - 770 - 780
790 - 800 - 810 - 820 - 830 - 840 - 850 - 860 - 870 - 880 - 890 - 900 - 910
920 - 930 - 940 - 950 - 960 - 970 - 980 - 990 - 1000

Notes: ____________________

Reg Reps: 10 - 20 - 30 - 40 - 50 - 60 - 70 - 80 - 90 - 100 - 110 - 120 - 130
140 - 150 - 160 - 170 - 180 - 190 - 200 - 210 - 220 - 230 - 240 - 250 - 260
270 - 280 - 290 - 300 - 310 - 320 - 330 - 340 - 350 - 360 - 370 - 380 - 390
400 - 410 - 420 - 430 - 440 - 450 - 460 - 470 - 480 - 490 - 500 - 510 - 520
530 - 540 - 550 - 560 - 570 - 580 - 590 - 600 - 610 - 620 - 630 - 640 - 650
660 - 670 - 680 - 690 - 700 - 710 - 720 - 730 - 740 - 750 - 760 - 770 - 780
790 - 800 - 810 - 820 - 830 - 840 - 850 - 860 - 870 - 880 - 890 - 900 - 910
920 - 930 - 940 - 950 - 960 - 970 - 980 - 990 - 1000

Notes: ____________________

End Date: ____________

Defense Against:_________________________ **Start Date:** __________

Notes:____

____

____

____

Slow Reps: 10 - 20 - 30 - 40 - 50 - 60 - 70 - 80 - 90 - 100 - 110 - 120 - 130
140 - 150 - 160 - 170 - 180 - 190 - 200 - 210 - 220 - 230 - 240 - 250 - 260
270 - 280 - 290 - 300 - 310 - 320 - 330 - 340 - 350 - 360 - 370 - 380 - 390
400 - 410 - 420 - 430 - 440 - 450 - 460 - 470 - 480 - 490 - 500 - 510 - 520
530 - 540 - 550 - 560 - 570 - 580 - 590 - 600 - 610 - 620 - 630 - 640 - 650
660 - 670 - 680 - 690 - 700 - 710 - 720 - 730 - 740 - 750 - 760 - 770 - 780
790 - 800 - 810 - 820 - 830 - 840 - 850 - 860 - 870 - 880 - 890 - 900 - 910
920 - 930 - 940 - 950 - 960 - 970 - 980 - 990 - 1000

Reg Reps: 10 - 20 - 30 - 40 - 50 - 60 - 70 - 80 - 90 - 100 - 110 - 120 - 130
140 - 150 - 160 - 170 - 180 - 190 - 200 - 210 - 220 - 230 - 240 - 250 - 260
270 - 280 - 290 - 300 - 310 - 320 - 330 - 340 - 350 - 360 - 370 - 380 - 390
400 - 410 - 420 - 430 - 440 - 450 - 460 - 470 - 480 - 490 - 500 - 510 - 520
530 - 540 - 550 - 560 - 570 - 580 - 590 - 600 - 610 - 620 - 630 - 640 - 650
660 - 670 - 680 - 690 - 700 - 710 - 720 - 730 - 740 - 750 - 760 - 770 - 780
790 - 800 - 810 - 820 - 830 - 840 - 850 - 860 - 870 - 880 - 890 - 900 - 910
920 - 930 - 940 - 950 - 960 - 970 - 980 - 990 - 1000

Notes: ____

____

____

____

Reg Reps: 10 - 20 - 30 - 40 - 50 - 60 - 70 - 80 - 90 - 100 - 110 - 120 - 130
140 - 150 - 160 - 170 - 180 - 190 - 200 - 210 - 220 - 230 - 240 - 250 - 260
270 - 280 - 290 - 300 - 310 - 320 - 330 - 340 - 350 - 360 - 370 - 380 - 390
400 - 410 - 420 - 430 - 440 - 450 - 460 - 470 - 480 - 490 - 500 - 510 - 520
530 - 540 - 550 - 560 - 570 - 580 - 590 - 600 - 610 - 620 - 630 - 640 - 650
660 - 670 - 680 - 690 - 700 - 710 - 720 - 730 - 740 - 750 - 760 - 770 - 780
790 - 800 - 810 - 820 - 830 - 840 - 850 - 860 - 870 - 880 - 890 - 900 - 910
920 - 930 - 940 - 950 - 960 - 970 - 980 - 990 - 1000

Notes: ____

____

____

____

Reg Reps: 10 - 20 - 30 - 40 - 50 - 60 - 70 - 80 - 90 - 100 - 110 - 120 - 130
140 - 150 - 160 - 170 - 180 - 190 - 200 - 210 - 220 - 230 - 240 - 250 - 260
270 - 280 - 290 - 300 - 310 - 320 - 330 - 340 - 350 - 360 - 370 - 380 - 390
400 - 410 - 420 - 430 - 440 - 450 - 460 - 470 - 480 - 490 - 500 - 510 - 520
530 - 540 - 550 - 560 - 570 - 580 - 590 - 600 - 610 - 620 - 630 - 640 - 650
660 - 670 - 680 - 690 - 700 - 710 - 720 - 730 - 740 - 750 - 760 - 770 - 780
790 - 800 - 810 - 820 - 830 - 840 - 850 - 860 - 870 - 880 - 890 - 900 - 910
920 - 930 - 940 - 950 - 960 - 970 - 980 - 990 - 1000

Notes: ____

____

____

____

Reg Reps: 10 - 20 - 30 - 40 - 50 - 60 - 70 - 80 - 90 - 100 - 110 - 120 - 130
140 - 150 - 160 - 170 - 180 - 190 - 200 - 210 - 220 - 230 - 240 - 250 - 260
270 - 280 - 290 - 300 - 310 - 320 - 330 - 340 - 350 - 360 - 370 - 380 - 390
400 - 410 - 420 - 430 - 440 - 450 - 460 - 470 - 480 - 490 - 500 - 510 - 520
530 - 540 - 550 - 560 - 570 - 580 - 590 - 600 - 610 - 620 - 630 - 640 - 650
660 - 670 - 680 - 690 - 700 - 710 - 720 - 730 - 740 - 750 - 760 - 770 - 780
790 - 800 - 810 - 820 - 830 - 840 - 850 - 860 - 870 - 880 - 890 - 900 - 910
920 - 930 - 940 - 950 - 960 - 970 - 980 - 990 - 1000

Notes: ____

____

____

____

Reg Reps: 10 - 20 - 30 - 40 - 50 - 60 - 70 - 80 - 90 - 100 - 110 - 120 - 130
140 - 150 - 160 - 170 - 180 - 190 - 200 - 210 - 220 - 230 - 240 - 250 - 260
270 - 280 - 290 - 300 - 310 - 320 - 330 - 340 - 350 - 360 - 370 - 380 - 390
400 - 410 - 420 - 430 - 440 - 450 - 460 - 470 - 480 - 490 - 500 - 510 - 520
530 - 540 - 550 - 560 - 570 - 580 - 590 - 600 - 610 - 620 - 630 - 640 - 650
660 - 670 - 680 - 690 - 700 - 710 - 720 - 730 - 740 - 750 - 760 - 770 - 780
790 - 800 - 810 - 820 - 830 - 840 - 850 - 860 - 870 - 880 - 890 - 900 - 910
920 - 930 - 940 - 950 - 960 - 970 - 980 - 990 - 1000

Notes: ____

____

____

______________________________ **End Date:** ____________

Defense Against:______________________________ **Start Date:** ____________

Notes:____

Slow Reps: 10 - 20 - 30 - 40 - 50 - 60 - 70 - 80 - 90 - 100 - 110 - 120 - 130
140 - 150 - 160 - 170 - 180 - 190 - 200 - 210 - 220 - 230 - 240 - 250 - 260
270 - 280 - 290 - 300 - 310 - 320 - 330 - 340 - 350 - 360 - 370 - 380 - 390
400 - 410 - 420 - 430 - 440 - 450 - 460 - 470 - 480 - 490 - 500 - 510 - 520
530 - 540 - 550 - 560 - 570 - 580 - 590 - 600 - 610 - 620 - 630 - 640 - 650
660 - 670 - 680 - 690 - 700 - 710 - 720 - 730 - 740 - 750 - 760 - 770 - 780
790 - 800 - 810 - 820 - 830 - 840 - 850 - 860 - 870 - 880 - 890 - 900 - 910
920 - 930 - 940 - 950 - 960 - 970 - 980 - 990 - 1000

Reg Reps: 10 - 20 - 30 - 40 - 50 - 60 - 70 - 80 - 90 - 100 - 110 - 120 - 130
140 - 150 - 160 - 170 - 180 - 190 - 200 - 210 - 220 - 230 - 240 - 250 - 260
270 - 280 - 290 - 300 - 310 - 320 - 330 - 340 - 350 - 360 - 370 - 380 - 390
400 - 410 - 420 - 430 - 440 - 450 - 460 - 470 - 480 - 490 - 500 - 510 - 520
530 - 540 - 550 - 560 - 570 - 580 - 590 - 600 - 610 - 620 - 630 - 640 - 650
660 - 670 - 680 - 690 - 700 - 710 - 720 - 730 - 740 - 750 - 760 - 770 - 780
790 - 800 - 810 - 820 - 830 - 840 - 850 - 860 - 870 - 880 - 890 - 900 - 910
920 - 930 - 940 - 950 - 960 - 970 - 980 - 990 - 1000

Notes: ____

Reg Reps: 10 - 20 - 30 - 40 - 50 - 60 - 70 - 80 - 90 - 100 - 110 - 120 - 130
140 - 150 - 160 - 170 - 180 - 190 - 200 - 210 - 220 - 230 - 240 - 250 - 260
270 - 280 - 290 - 300 - 310 - 320 - 330 - 340 - 350 - 360 - 370 - 380 - 390
400 - 410 - 420 - 430 - 440 - 450 - 460 - 470 - 480 - 490 - 500 - 510 - 520
530 - 540 - 550 - 560 - 570 - 580 - 590 - 600 - 610 - 620 - 630 - 640 - 650
660 - 670 - 680 - 690 - 700 - 710 - 720 - 730 - 740 - 750 - 760 - 770 - 780
790 - 800 - 810 - 820 - 830 - 840 - 850 - 860 - 870 - 880 - 890 - 900 - 910
920 - 930 - 940 - 950 - 960 - 970 - 980 - 990 - 1000

Notes: ____

Reg Reps: 10 - 20 - 30 - 40 - 50 - 60 - 70 - 80 - 90 - 100 - 110 - 120 - 130
140 - 150 - 160 - 170 - 180 - 190 - 200 - 210 - 220 - 230 - 240 - 250 - 260
270 - 280 - 290 - 300 - 310 - 320 - 330 - 340 - 350 - 360 - 370 - 380 - 390
400 - 410 - 420 - 430 - 440 - 450 - 460 - 470 - 480 - 490 - 500 - 510 - 520
530 - 540 - 550 - 560 - 570 - 580 - 590 - 600 - 610 - 620 - 630 - 640 - 650
660 - 670 - 680 - 690 - 700 - 710 - 720 - 730 - 740 - 750 - 760 - 770 - 780
790 - 800 - 810 - 820 - 830 - 840 - 850 - 860 - 870 - 880 - 890 - 900 - 910
920 - 930 - 940 - 950 - 960 - 970 - 980 - 990 - 1000

Notes: ______________________________

____

____

____

Reg Reps: 10 - 20 - 30 - 40 - 50 - 60 - 70 - 80 - 90 - 100 - 110 - 120 - 130
140 - 150 - 160 - 170 - 180 - 190 - 200 - 210 - 220 - 230 - 240 - 250 - 260
270 - 280 - 290 - 300 - 310 - 320 - 330 - 340 - 350 - 360 - 370 - 380 - 390
400 - 410 - 420 - 430 - 440 - 450 - 460 - 470 - 480 - 490 - 500 - 510 - 520
530 - 540 - 550 - 560 - 570 - 580 - 590 - 600 - 610 - 620 - 630 - 640 - 650
660 - 670 - 680 - 690 - 700 - 710 - 720 - 730 - 740 - 750 - 760 - 770 - 780
790 - 800 - 810 - 820 - 830 - 840 - 850 - 860 - 870 - 880 - 890 - 900 - 910
920 - 930 - 940 - 950 - 960 - 970 - 980 - 990 - 1000

Notes: ______________________________

____

____

____

Reg Reps: 10 - 20 - 30 - 40 - 50 - 60 - 70 - 80 - 90 - 100 - 110 - 120 - 130
140 - 150 - 160 - 170 - 180 - 190 - 200 - 210 - 220 - 230 - 240 - 250 - 260
270 - 280 - 290 - 300 - 310 - 320 - 330 - 340 - 350 - 360 - 370 - 380 - 390
400 - 410 - 420 - 430 - 440 - 450 - 460 - 470 - 480 - 490 - 500 - 510 - 520
530 - 540 - 550 - 560 - 570 - 580 - 590 - 600 - 610 - 620 - 630 - 640 - 650
660 - 670 - 680 - 690 - 700 - 710 - 720 - 730 - 740 - 750 - 760 - 770 - 780
790 - 800 - 810 - 820 - 830 - 840 - 850 - 860 - 870 - 880 - 890 - 900 - 910
920 - 930 - 940 - 950 - 960 - 970 - 980 - 990 - 1000

Notes: ______________________________

____

____

______________________________ **End Date:** __________

Defense Against: ______________________ **Start Date:** __________

Notes: ____

____

____

____

Slow Reps: 10 - 20 - 30 - 40 - 50 - 60 - 70 - 80 - 90 - 100 - 110 - 120 - 130
140 - 150 - 160 - 170 - 180 - 190 - 200 - 210 - 220 - 230 - 240 - 250 - 260
270 - 280 - 290 - 300 - 310 - 320 - 330 - 340 - 350 - 360 - 370 - 380 - 390
400 - 410 - 420 - 430 - 440 - 450 - 460 - 470 - 480 - 490 - 500 - 510 - 520
530 - 540 - 550 - 560 - 570 - 580 - 590 - 600 - 610 - 620 - 630 - 640 - 650
660 - 670 - 680 - 690 - 700 - 710 - 720 - 730 - 740 - 750 - 760 - 770 - 780
790 - 800 - 810 - 820 - 830 - 840 - 850 - 860 - 870 - 880 - 890 - 900 - 910
920 - 930 - 940 - 950 - 960 - 970 - 980 - 990 - 1000

Reg Reps: 10 - 20 - 30 - 40 - 50 - 60 - 70 - 80 - 90 - 100 - 110 - 120 - 130
140 - 150 - 160 - 170 - 180 - 190 - 200 - 210 - 220 - 230 - 240 - 250 - 260
270 - 280 - 290 - 300 - 310 - 320 - 330 - 340 - 350 - 360 - 370 - 380 - 390
400 - 410 - 420 - 430 - 440 - 450 - 460 - 470 - 480 - 490 - 500 - 510 - 520
530 - 540 - 550 - 560 - 570 - 580 - 590 - 600 - 610 - 620 - 630 - 640 - 650
660 - 670 - 680 - 690 - 700 - 710 - 720 - 730 - 740 - 750 - 760 - 770 - 780
790 - 800 - 810 - 820 - 830 - 840 - 850 - 860 - 870 - 880 - 890 - 900 - 910
920 - 930 - 940 - 950 - 960 - 970 - 980 - 990 - 1000

Notes: ____

____

____

____

Reg Reps: 10 - 20 - 30 - 40 - 50 - 60 - 70 - 80 - 90 - 100 - 110 - 120 - 130
140 - 150 - 160 - 170 - 180 - 190 - 200 - 210 - 220 - 230 - 240 - 250 - 260
270 - 280 - 290 - 300 - 310 - 320 - 330 - 340 - 350 - 360 - 370 - 380 - 390
400 - 410 - 420 - 430 - 440 - 450 - 460 - 470 - 480 - 490 - 500 - 510 - 520
530 - 540 - 550 - 560 - 570 - 580 - 590 - 600 - 610 - 620 - 630 - 640 - 650
660 - 670 - 680 - 690 - 700 - 710 - 720 - 730 - 740 - 750 - 760 - 770 - 780
790 - 800 - 810 - 820 - 830 - 840 - 850 - 860 - 870 - 880 - 890 - 900 - 910
920 - 930 - 940 - 950 - 960 - 970 - 980 - 990 - 1000

Notes: ____

____

____

____

Reg Reps: 10 - 20 - 30 - 40 - 50 - 60 - 70 - 80 - 90 - 100 - 110 - 120 - 130 - 140 - 150 - 160 - 170 - 180 - 190 - 200 - 210 - 220 - 230 - 240 - 250 - 260 - 270 - 280 - 290 - 300 - 310 - 320 - 330 - 340 - 350 - 360 - 370 - 380 - 390 - 400 - 410 - 420 - 430 - 440 - 450 - 460 - 470 - 480 - 490 - 500 - 510 - 520 - 530 - 540 - 550 - 560 - 570 - 580 - 590 - 600 - 610 - 620 - 630 - 640 - 650 - 660 - 670 - 680 - 690 - 700 - 710 - 720 - 730 - 740 - 750 - 760 - 770 - 780 - 790 - 800 - 810 - 820 - 830 - 840 - 850 - 860 - 870 - 880 - 890 - 900 - 910 - 920 - 930 - 940 - 950 - 960 - 970 - 980 - 990 - 1000

Notes: ______________________________

____

____

____

Reg Reps: 10 - 20 - 30 - 40 - 50 - 60 - 70 - 80 - 90 - 100 - 110 - 120 - 130 - 140 - 150 - 160 - 170 - 180 - 190 - 200 - 210 - 220 - 230 - 240 - 250 - 260 - 270 - 280 - 290 - 300 - 310 - 320 - 330 - 340 - 350 - 360 - 370 - 380 - 390 - 400 - 410 - 420 - 430 - 440 - 450 - 460 - 470 - 480 - 490 - 500 - 510 - 520 - 530 - 540 - 550 - 560 - 570 - 580 - 590 - 600 - 610 - 620 - 630 - 640 - 650 - 660 - 670 - 680 - 690 - 700 - 710 - 720 - 730 - 740 - 750 - 760 - 770 - 780 - 790 - 800 - 810 - 820 - 830 - 840 - 850 - 860 - 870 - 880 - 890 - 900 - 910 - 920 - 930 - 940 - 950 - 960 - 970 - 980 - 990 - 1000

Notes: ______________________________

____

____

____

Reg Reps: 10 - 20 - 30 - 40 - 50 - 60 - 70 - 80 - 90 - 100 - 110 - 120 - 130 - 140 - 150 - 160 - 170 - 180 - 190 - 200 - 210 - 220 - 230 - 240 - 250 - 260 - 270 - 280 - 290 - 300 - 310 - 320 - 330 - 340 - 350 - 360 - 370 - 380 - 390 - 400 - 410 - 420 - 430 - 440 - 450 - 460 - 470 - 480 - 490 - 500 - 510 - 520 - 530 - 540 - 550 - 560 - 570 - 580 - 590 - 600 - 610 - 620 - 630 - 640 - 650 - 660 - 670 - 680 - 690 - 700 - 710 - 720 - 730 - 740 - 750 - 760 - 770 - 780 - 790 - 800 - 810 - 820 - 830 - 840 - 850 - 860 - 870 - 880 - 890 - 900 - 910 - 920 - 930 - 940 - 950 - 960 - 970 - 980 - 990 - 1000

Notes: ______________________________

____

____

______________________________ **End Date:** ______________

Defense Against:_________________________ **Start Date:** __________

Notes:____

____

____

____

Slow Reps: 10 - 20 - 30 - 40 - 50 - 60 - 70 - 80 - 90 - 100 - 110 - 120 - 130
140 - 150 - 160 - 170 - 180 - 190 - 200 - 210 - 220 - 230 - 240 - 250 - 260
270 - 280 - 290 - 300 - 310 - 320 - 330 - 340 - 350 - 360 - 370 - 380 - 390
400 - 410 - 420 - 430 - 440 - 450 - 460 - 470 - 480 - 490 - 500 - 510 - 520
530 - 540 - 550 - 560 - 570 - 580 - 590 - 600 - 610 - 620 - 630 - 640 - 650
660 - 670 - 680 - 690 - 700 - 710 - 720 - 730 - 740 - 750 - 760 - 770 - 780
790 - 800 - 810 - 820 - 830 - 840 - 850 - 860 - 870 - 880 - 890 - 900 - 910
920 - 930 - 940 - 950 - 960 - 970 - 980 - 990 - 1000

Reg Reps: 10 - 20 - 30 - 40 - 50 - 60 - 70 - 80 - 90 - 100 - 110 - 120 - 130
140 - 150 - 160 - 170 - 180 - 190 - 200 - 210 - 220 - 230 - 240 - 250 - 260
270 - 280 - 290 - 300 - 310 - 320 - 330 - 340 - 350 - 360 - 370 - 380 - 390
400 - 410 - 420 - 430 - 440 - 450 - 460 - 470 - 480 - 490 - 500 - 510 - 520
530 - 540 - 550 - 560 - 570 - 580 - 590 - 600 - 610 - 620 - 630 - 640 - 650
660 - 670 - 680 - 690 - 700 - 710 - 720 - 730 - 740 - 750 - 760 - 770 - 780
790 - 800 - 810 - 820 - 830 - 840 - 850 - 860 - 870 - 880 - 890 - 900 - 910
920 - 930 - 940 - 950 - 960 - 970 - 980 - 990 - 1000

Notes: ____

____

____

____

Reg Reps: 10 - 20 - 30 - 40 - 50 - 60 - 70 - 80 - 90 - 100 - 110 - 120 - 130
140 - 150 - 160 - 170 - 180 - 190 - 200 - 210 - 220 - 230 - 240 - 250 - 260
270 - 280 - 290 - 300 - 310 - 320 - 330 - 340 - 350 - 360 - 370 - 380 - 390
400 - 410 - 420 - 430 - 440 - 450 - 460 - 470 - 480 - 490 - 500 - 510 - 520
530 - 540 - 550 - 560 - 570 - 580 - 590 - 600 - 610 - 620 - 630 - 640 - 650
660 - 670 - 680 - 690 - 700 - 710 - 720 - 730 - 740 - 750 - 760 - 770 - 780
790 - 800 - 810 - 820 - 830 - 840 - 850 - 860 - 870 - 880 - 890 - 900 - 910
920 - 930 - 940 - 950 - 960 - 970 - 980 - 990 - 1000

Notes: ____

____

____

____

Reg Reps: 10 - 20 - 30 - 40 - 50 - 60 - 70 - 80 - 90 - 100 - 110 - 120 - 130 - 140 - 150 - 160 - 170 - 180 - 190 - 200 - 210 - 220 - 230 - 240 - 250 - 260 - 270 - 280 - 290 - 300 - 310 - 320 - 330 - 340 - 350 - 360 - 370 - 380 - 390 - 400 - 410 - 420 - 430 - 440 - 450 - 460 - 470 - 480 - 490 - 500 - 510 - 520 - 530 - 540 - 550 - 560 - 570 - 580 - 590 - 600 - 610 - 620 - 630 - 640 - 650 - 660 - 670 - 680 - 690 - 700 - 710 - 720 - 730 - 740 - 750 - 760 - 770 - 780 - 790 - 800 - 810 - 820 - 830 - 840 - 850 - 860 - 870 - 880 - 890 - 900 - 910 - 920 - 930 - 940 - 950 - 960 - 970 - 980 - 990 - 1000

Notes: ______________________________

____

____

____

Reg Reps: 10 - 20 - 30 - 40 - 50 - 60 - 70 - 80 - 90 - 100 - 110 - 120 - 130 - 140 - 150 - 160 - 170 - 180 - 190 - 200 - 210 - 220 - 230 - 240 - 250 - 260 - 270 - 280 - 290 - 300 - 310 - 320 - 330 - 340 - 350 - 360 - 370 - 380 - 390 - 400 - 410 - 420 - 430 - 440 - 450 - 460 - 470 - 480 - 490 - 500 - 510 - 520 - 530 - 540 - 550 - 560 - 570 - 580 - 590 - 600 - 610 - 620 - 630 - 640 - 650 - 660 - 670 - 680 - 690 - 700 - 710 - 720 - 730 - 740 - 750 - 760 - 770 - 780 - 790 - 800 - 810 - 820 - 830 - 840 - 850 - 860 - 870 - 880 - 890 - 900 - 910 - 920 - 930 - 940 - 950 - 960 - 970 - 980 - 990 - 1000

Notes: ______________________________

____

____

____

Reg Reps: 10 - 20 - 30 - 40 - 50 - 60 - 70 - 80 - 90 - 100 - 110 - 120 - 130 - 140 - 150 - 160 - 170 - 180 - 190 - 200 - 210 - 220 - 230 - 240 - 250 - 260 - 270 - 280 - 290 - 300 - 310 - 320 - 330 - 340 - 350 - 360 - 370 - 380 - 390 - 400 - 410 - 420 - 430 - 440 - 450 - 460 - 470 - 480 - 490 - 500 - 510 - 520 - 530 - 540 - 550 - 560 - 570 - 580 - 590 - 600 - 610 - 620 - 630 - 640 - 650 - 660 - 670 - 680 - 690 - 700 - 710 - 720 - 730 - 740 - 750 - 760 - 770 - 780 - 790 - 800 - 810 - 820 - 830 - 840 - 850 - 860 - 870 - 880 - 890 - 900 - 910 - 920 - 930 - 940 - 950 - 960 - 970 - 980 - 990 - 1000

Notes: ______________________________

____

____

______________________________ **End Date:** ____________

Defense Against:__________________________ **Start Date:** __________

Notes:____

Slow Reps: 10 - 20 - 30 - 40 - 50 - 60 - 70 - 80 - 90 - 100 - 110 - 120 - 130
140 - 150 - 160 - 170 - 180 - 190 - 200 - 210 - 220 - 230 - 240 - 250 - 260
270 - 280 - 290 - 300 - 310 - 320 - 330 - 340 - 350 - 360 - 370 - 380 - 390
400 - 410 - 420 - 430 - 440 - 450 - 460 - 470 - 480 - 490 - 500 - 510 - 520
530 - 540 - 550 - 560 - 570 - 580 - 590 - 600 - 610 - 620 - 630 - 640 - 650
660 - 670 - 680 - 690 - 700 - 710 - 720 - 730 - 740 - 750 - 760 - 770 - 780
790 - 800 - 810 - 820 - 830 - 840 - 850 - 860 - 870 - 880 - 890 - 900 - 910
920 - 930 - 940 - 950 - 960 - 970 - 980 - 990 - 1000

Reg Reps: 10 - 20 - 30 - 40 - 50 - 60 - 70 - 80 - 90 - 100 - 110 - 120 - 130
140 - 150 - 160 - 170 - 180 - 190 - 200 - 210 - 220 - 230 - 240 - 250 - 260
270 - 280 - 290 - 300 - 310 - 320 - 330 - 340 - 350 - 360 - 370 - 380 - 390
400 - 410 - 420 - 430 - 440 - 450 - 460 - 470 - 480 - 490 - 500 - 510 - 520
530 - 540 - 550 - 560 - 570 - 580 - 590 - 600 - 610 - 620 - 630 - 640 - 650
660 - 670 - 680 - 690 - 700 - 710 - 720 - 730 - 740 - 750 - 760 - 770 - 780
790 - 800 - 810 - 820 - 830 - 840 - 850 - 860 - 870 - 880 - 890 - 900 - 910
920 - 930 - 940 - 950 - 960 - 970 - 980 - 990 - 1000

Notes: ____

Reg Reps: 10 - 20 - 30 - 40 - 50 - 60 - 70 - 80 - 90 - 100 - 110 - 120 - 130
140 - 150 - 160 - 170 - 180 - 190 - 200 - 210 - 220 - 230 - 240 - 250 - 260
270 - 280 - 290 - 300 - 310 - 320 - 330 - 340 - 350 - 360 - 370 - 380 - 390
400 - 410 - 420 - 430 - 440 - 450 - 460 - 470 - 480 - 490 - 500 - 510 - 520
530 - 540 - 550 - 560 - 570 - 580 - 590 - 600 - 610 - 620 - 630 - 640 - 650
660 - 670 - 680 - 690 - 700 - 710 - 720 - 730 - 740 - 750 - 760 - 770 - 780
790 - 800 - 810 - 820 - 830 - 840 - 850 - 860 - 870 - 880 - 890 - 900 - 910
920 - 930 - 940 - 950 - 960 - 970 - 980 - 990 - 1000

Notes: ____

Reg Reps: 10 - 20 - 30 - 40 - 50 - 60 - 70 - 80 - 90 - 100 - 110 - 120 - 130
140 - 150 - 160 - 170 - 180 - 190 - 200 - 210 - 220 - 230 - 240 - 250 - 260
270 - 280 - 290 - 300 - 310 - 320 - 330 - 340 - 350 - 360 - 370 - 380 - 390
400 - 410 - 420 - 430 - 440 - 450 - 460 - 470 - 480 - 490 - 500 - 510 - 520
530 - 540 - 550 - 560 - 570 - 580 - 590 - 600 - 610 - 620 - 630 - 640 - 650
660 - 670 - 680 - 690 - 700 - 710 - 720 - 730 - 740 - 750 - 760 - 770 - 780
790 - 800 - 810 - 820 - 830 - 840 - 850 - 860 - 870 - 880 - 890 - 900 - 910
920 - 930 - 940 - 950 - 960 - 970 - 980 - 990 - 1000

Notes: ____

____

____

____

Reg Reps: 10 - 20 - 30 - 40 - 50 - 60 - 70 - 80 - 90 - 100 - 110 - 120 - 130
140 - 150 - 160 - 170 - 180 - 190 - 200 - 210 - 220 - 230 - 240 - 250 - 260
270 - 280 - 290 - 300 - 310 - 320 - 330 - 340 - 350 - 360 - 370 - 380 - 390
400 - 410 - 420 - 430 - 440 - 450 - 460 - 470 - 480 - 490 - 500 - 510 - 520
530 - 540 - 550 - 560 - 570 - 580 - 590 - 600 - 610 - 620 - 630 - 640 - 650
660 - 670 - 680 - 690 - 700 - 710 - 720 - 730 - 740 - 750 - 760 - 770 - 780
790 - 800 - 810 - 820 - 830 - 840 - 850 - 860 - 870 - 880 - 890 - 900 - 910
920 - 930 - 940 - 950 - 960 - 970 - 980 - 990 - 1000

Notes: ____

____

____

____

Reg Reps: 10 - 20 - 30 - 40 - 50 - 60 - 70 - 80 - 90 - 100 - 110 - 120 - 130
140 - 150 - 160 - 170 - 180 - 190 - 200 - 210 - 220 - 230 - 240 - 250 - 260
270 - 280 - 290 - 300 - 310 - 320 - 330 - 340 - 350 - 360 - 370 - 380 - 390
400 - 410 - 420 - 430 - 440 - 450 - 460 - 470 - 480 - 490 - 500 - 510 - 520
530 - 540 - 550 - 560 - 570 - 580 - 590 - 600 - 610 - 620 - 630 - 640 - 650
660 - 670 - 680 - 690 - 700 - 710 - 720 - 730 - 740 - 750 - 760 - 770 - 780
790 - 800 - 810 - 820 - 830 - 840 - 850 - 860 - 870 - 880 - 890 - 900 - 910
920 - 930 - 940 - 950 - 960 - 970 - 980 - 990 - 1000

Notes: ____

____

____

______________________________ **End Date:** ____________

Defense Against:__________________________ **Start Date:** __________

Notes:____

Slow Reps: 10 - 20 - 30 - 40 - 50 - 60 - 70 - 80 - 90 - 100 - 110 - 120 - 130
140 - 150 - 160 - 170 - 180 - 190 - 200 - 210 - 220 - 230 - 240 - 250 - 260
270 - 280 - 290 - 300 - 310 - 320 - 330 - 340 - 350 - 360 - 370 - 380 - 390
400 - 410 - 420 - 430 - 440 - 450 - 460 - 470 - 480 - 490 - 500 - 510 - 520
530 - 540 - 550 - 560 - 570 - 580 - 590 - 600 - 610 - 620 - 630 - 640 - 650
660 - 670 - 680 - 690 - 700 - 710 - 720 - 730 - 740 - 750 - 760 - 770 - 780
790 - 800 - 810 - 820 - 830 - 840 - 850 - 860 - 870 - 880 - 890 - 900 - 910
920 - 930 - 940 - 950 - 960 - 970 - 980 - 990 - 1000

Reg Reps: 10 - 20 - 30 - 40 - 50 - 60 - 70 - 80 - 90 - 100 - 110 - 120 - 130
140 - 150 - 160 - 170 - 180 - 190 - 200 - 210 - 220 - 230 - 240 - 250 - 260
270 - 280 - 290 - 300 - 310 - 320 - 330 - 340 - 350 - 360 - 370 - 380 - 390
400 - 410 - 420 - 430 - 440 - 450 - 460 - 470 - 480 - 490 - 500 - 510 - 520
530 - 540 - 550 - 560 - 570 - 580 - 590 - 600 - 610 - 620 - 630 - 640 - 650
660 - 670 - 680 - 690 - 700 - 710 - 720 - 730 - 740 - 750 - 760 - 770 - 780
790 - 800 - 810 - 820 - 830 - 840 - 850 - 860 - 870 - 880 - 890 - 900 - 910
920 - 930 - 940 - 950 - 960 - 970 - 980 - 990 - 1000

Notes: ____

Reg Reps: 10 - 20 - 30 - 40 - 50 - 60 - 70 - 80 - 90 - 100 - 110 - 120 - 130
140 - 150 - 160 - 170 - 180 - 190 - 200 - 210 - 220 - 230 - 240 - 250 - 260
270 - 280 - 290 - 300 - 310 - 320 - 330 - 340 - 350 - 360 - 370 - 380 - 390
400 - 410 - 420 - 430 - 440 - 450 - 460 - 470 - 480 - 490 - 500 - 510 - 520
530 - 540 - 550 - 560 - 570 - 580 - 590 - 600 - 610 - 620 - 630 - 640 - 650
660 - 670 - 680 - 690 - 700 - 710 - 720 - 730 - 740 - 750 - 760 - 770 - 780
790 - 800 - 810 - 820 - 830 - 840 - 850 - 860 - 870 - 880 - 890 - 900 - 910
920 - 930 - 940 - 950 - 960 - 970 - 980 - 990 - 1000

Notes: ____

Reg Reps: 10 - 20 - 30 - 40 - 50 - 60 - 70 - 80 - 90 - 100 - 110 - 120 - 130
140 - 150 - 160 - 170 - 180 - 190 - 200 - 210 - 220 - 230 - 240 - 250 - 260
270 - 280 - 290 - 300 - 310 - 320 - 330 - 340 - 350 - 360 - 370 - 380 - 390
400 - 410 - 420 - 430 - 440 - 450 - 460 - 470 - 480 - 490 - 500 - 510 - 520
530 - 540 - 550 - 560 - 570 - 580 - 590 - 600 - 610 - 620 - 630 - 640 - 650
660 - 670 - 680 - 690 - 700 - 710 - 720 - 730 - 740 - 750 - 760 - 770 - 780
790 - 800 - 810 - 820 - 830 - 840 - 850 - 860 - 870 - 880 - 890 - 900 - 910
920 - 930 - 940 - 950 - 960 - 970 - 980 - 990 - 1000

Notes: ____

____

____

____

Reg Reps: 10 - 20 - 30 - 40 - 50 - 60 - 70 - 80 - 90 - 100 - 110 - 120 - 130
140 - 150 - 160 - 170 - 180 - 190 - 200 - 210 - 220 - 230 - 240 - 250 - 260
270 - 280 - 290 - 300 - 310 - 320 - 330 - 340 - 350 - 360 - 370 - 380 - 390
400 - 410 - 420 - 430 - 440 - 450 - 460 - 470 - 480 - 490 - 500 - 510 - 520
530 - 540 - 550 - 560 - 570 - 580 - 590 - 600 - 610 - 620 - 630 - 640 - 650
660 - 670 - 680 - 690 - 700 - 710 - 720 - 730 - 740 - 750 - 760 - 770 - 780
790 - 800 - 810 - 820 - 830 - 840 - 850 - 860 - 870 - 880 - 890 - 900 - 910
920 - 930 - 940 - 950 - 960 - 970 - 980 - 990 - 1000

Notes: ____

____

____

____

Reg Reps: 10 - 20 - 30 - 40 - 50 - 60 - 70 - 80 - 90 - 100 - 110 - 120 - 130
140 - 150 - 160 - 170 - 180 - 190 - 200 - 210 - 220 - 230 - 240 - 250 - 260
270 - 280 - 290 - 300 - 310 - 320 - 330 - 340 - 350 - 360 - 370 - 380 - 390
400 - 410 - 420 - 430 - 440 - 450 - 460 - 470 - 480 - 490 - 500 - 510 - 520
530 - 540 - 550 - 560 - 570 - 580 - 590 - 600 - 610 - 620 - 630 - 640 - 650
660 - 670 - 680 - 690 - 700 - 710 - 720 - 730 - 740 - 750 - 760 - 770 - 780
790 - 800 - 810 - 820 - 830 - 840 - 850 - 860 - 870 - 880 - 890 - 900 - 910
920 - 930 - 940 - 950 - 960 - 970 - 980 - 990 - 1000

Notes: ____

____

____

______________________________ **End Date:** ____________

Defense Against:_________________________ **Start Date:** __________

Notes:____

____

____

____

Slow Reps: 10 - 20 - 30 - 40 - 50 - 60 - 70 - 80 - 90 - 100 - 110 - 120 - 130
140 - 150 - 160 - 170 - 180 - 190 - 200 - 210 - 220 - 230 - 240 - 250 - 260
270 - 280 - 290 - 300 - 310 - 320 - 330 - 340 - 350 - 360 - 370 - 380 - 390
400 - 410 - 420 - 430 - 440 - 450 - 460 - 470 - 480 - 490 - 500 - 510 - 520
530 - 540 - 550 - 560 - 570 - 580 - 590 - 600 - 610 - 620 - 630 - 640 - 650
660 - 670 - 680 - 690 - 700 - 710 - 720 - 730 - 740 - 750 - 760 - 770 - 780
790 - 800 - 810 - 820 - 830 - 840 - 850 - 860 - 870 - 880 - 890 - 900 - 910
920 - 930 - 940 - 950 - 960 - 970 - 980 - 990 - 1000

Reg Reps: 10 - 20 - 30 - 40 - 50 - 60 - 70 - 80 - 90 - 100 - 110 - 120 - 130
140 - 150 - 160 - 170 - 180 - 190 - 200 - 210 - 220 - 230 - 240 - 250 - 260
270 - 280 - 290 - 300 - 310 - 320 - 330 - 340 - 350 - 360 - 370 - 380 - 390
400 - 410 - 420 - 430 - 440 - 450 - 460 - 470 - 480 - 490 - 500 - 510 - 520
530 - 540 - 550 - 560 - 570 - 580 - 590 - 600 - 610 - 620 - 630 - 640 - 650
660 - 670 - 680 - 690 - 700 - 710 - 720 - 730 - 740 - 750 - 760 - 770 - 780
790 - 800 - 810 - 820 - 830 - 840 - 850 - 860 - 870 - 880 - 890 - 900 - 910
920 - 930 - 940 - 950 - 960 - 970 - 980 - 990 - 1000

Notes: ____

____

____

____

Reg Reps: 10 - 20 - 30 - 40 - 50 - 60 - 70 - 80 - 90 - 100 - 110 - 120 - 130
140 - 150 - 160 - 170 - 180 - 190 - 200 - 210 - 220 - 230 - 240 - 250 - 260
270 - 280 - 290 - 300 - 310 - 320 - 330 - 340 - 350 - 360 - 370 - 380 - 390
400 - 410 - 420 - 430 - 440 - 450 - 460 - 470 - 480 - 490 - 500 - 510 - 520
530 - 540 - 550 - 560 - 570 - 580 - 590 - 600 - 610 - 620 - 630 - 640 - 650
660 - 670 - 680 - 690 - 700 - 710 - 720 - 730 - 740 - 750 - 760 - 770 - 780
790 - 800 - 810 - 820 - 830 - 840 - 850 - 860 - 870 - 880 - 890 - 900 - 910
920 - 930 - 940 - 950 - 960 - 970 - 980 - 990 - 1000

Notes: ____

____

____

____

Reg Reps: 10 - 20 - 30 - 40 - 50 - 60 - 70 - 80 - 90 - 100 - 110 - 120 - 130 140 - 150 - 160 - 170 - 180 - 190 - 200 - 210 - 220 - 230 - 240 - 250 - 260 270 - 280 - 290 - 300 - 310 - 320 - 330 - 340 - 350 - 360 - 370 - 380 - 390 400 - 410 - 420 - 430 - 440 - 450 - 460 - 470 - 480 - 490 - 500 - 510 - 520 530 - 540 - 550 - 560 - 570 - 580 - 590 - 600 - 610 - 620 - 630 - 640 - 650 660 - 670 - 680 - 690 - 700 - 710 - 720 - 730 - 740 - 750 - 760 - 770 - 780 790 - 800 - 810 - 820 - 830 - 840 - 850 - 860 - 870 - 880 - 890 - 900 - 910 920 - 930 - 940 - 950 - 960 - 970 - 980 - 990 - 1000

Notes: ______________________________

____

____

____

Reg Reps: 10 - 20 - 30 - 40 - 50 - 60 - 70 - 80 - 90 - 100 - 110 - 120 - 130 140 - 150 - 160 - 170 - 180 - 190 - 200 - 210 - 220 - 230 - 240 - 250 - 260 270 - 280 - 290 - 300 - 310 - 320 - 330 - 340 - 350 - 360 - 370 - 380 - 390 400 - 410 - 420 - 430 - 440 - 450 - 460 - 470 - 480 - 490 - 500 - 510 - 520 530 - 540 - 550 - 560 - 570 - 580 - 590 - 600 - 610 - 620 - 630 - 640 - 650 660 - 670 - 680 - 690 - 700 - 710 - 720 - 730 - 740 - 750 - 760 - 770 - 780 790 - 800 - 810 - 820 - 830 - 840 - 850 - 860 - 870 - 880 - 890 - 900 - 910 920 - 930 - 940 - 950 - 960 - 970 - 980 - 990 - 1000

Notes: ______________________________

____

____

____

Reg Reps: 10 - 20 - 30 - 40 - 50 - 60 - 70 - 80 - 90 - 100 - 110 - 120 - 130 140 - 150 - 160 - 170 - 180 - 190 - 200 - 210 - 220 - 230 - 240 - 250 - 260 270 - 280 - 290 - 300 - 310 - 320 - 330 - 340 - 350 - 360 - 370 - 380 - 390 400 - 410 - 420 - 430 - 440 - 450 - 460 - 470 - 480 - 490 - 500 - 510 - 520 530 - 540 - 550 - 560 - 570 - 580 - 590 - 600 - 610 - 620 - 630 - 640 - 650 660 - 670 - 680 - 690 - 700 - 710 - 720 - 730 - 740 - 750 - 760 - 770 - 780 790 - 800 - 810 - 820 - 830 - 840 - 850 - 860 - 870 - 880 - 890 - 900 - 910 920 - 930 - 940 - 950 - 960 - 970 - 980 - 990 - 1000

Notes: ______________________________

____

____

______________________________ **End Date:** __________

Defense Against:______________________________ **Start Date:** ___________

Notes:____

____

____

____

Slow Reps: 10 - 20 - 30 - 40 - 50 - 60 - 70 - 80 - 90 - 100 - 110 - 120 - 130
140 - 150 - 160 - 170 - 180 - 190 - 200 - 210 - 220 - 230 - 240 - 250 - 260
270 - 280 - 290 - 300 - 310 - 320 - 330 - 340 - 350 - 360 - 370 - 380 - 390
400 - 410 - 420 - 430 - 440 - 450 - 460 - 470 - 480 - 490 - 500 - 510 - 520
530 - 540 - 550 - 560 - 570 - 580 - 590 - 600 - 610 - 620 - 630 - 640 - 650
660 - 670 - 680 - 690 - 700 - 710 - 720 - 730 - 740 - 750 - 760 - 770 - 780
790 - 800 - 810 - 820 - 830 - 840 - 850 - 860 - 870 - 880 - 890 - 900 - 910
920 - 930 - 940 - 950 - 960 - 970 - 980 - 990 - 1000

Reg Reps: 10 - 20 - 30 - 40 - 50 - 60 - 70 - 80 - 90 - 100 - 110 - 120 - 130
140 - 150 - 160 - 170 - 180 - 190 - 200 - 210 - 220 - 230 - 240 - 250 - 260
270 - 280 - 290 - 300 - 310 - 320 - 330 - 340 - 350 - 360 - 370 - 380 - 390
400 - 410 - 420 - 430 - 440 - 450 - 460 - 470 - 480 - 490 - 500 - 510 - 520
530 - 540 - 550 - 560 - 570 - 580 - 590 - 600 - 610 - 620 - 630 - 640 - 650
660 - 670 - 680 - 690 - 700 - 710 - 720 - 730 - 740 - 750 - 760 - 770 - 780
790 - 800 - 810 - 820 - 830 - 840 - 850 - 860 - 870 - 880 - 890 - 900 - 910
920 - 930 - 940 - 950 - 960 - 970 - 980 - 990 - 1000

Notes: ____

____

____

____

Reg Reps: 10 - 20 - 30 - 40 - 50 - 60 - 70 - 80 - 90 - 100 - 110 - 120 - 130
140 - 150 - 160 - 170 - 180 - 190 - 200 - 210 - 220 - 230 - 240 - 250 - 260
270 - 280 - 290 - 300 - 310 - 320 - 330 - 340 - 350 - 360 - 370 - 380 - 390
400 - 410 - 420 - 430 - 440 - 450 - 460 - 470 - 480 - 490 - 500 - 510 - 520
530 - 540 - 550 - 560 - 570 - 580 - 590 - 600 - 610 - 620 - 630 - 640 - 650
660 - 670 - 680 - 690 - 700 - 710 - 720 - 730 - 740 - 750 - 760 - 770 - 780
790 - 800 - 810 - 820 - 830 - 840 - 850 - 860 - 870 - 880 - 890 - 900 - 910
920 - 930 - 940 - 950 - 960 - 970 - 980 - 990 - 1000

Notes: ____

____

____

____

Reg Reps: 10 - 20 - 30 - 40 - 50 - 60 - 70 - 80 - 90 - 100 - 110 - 120 - 130
140 - 150 - 160 - 170 - 180 - 190 - 200 - 210 - 220 - 230 - 240 - 250 - 260
270 - 280 - 290 - 300 - 310 - 320 - 330 - 340 - 350 - 360 - 370 - 380 - 390
400 - 410 - 420 - 430 - 440 - 450 - 460 - 470 - 480 - 490 - 500 - 510 - 520
530 - 540 - 550 - 560 - 570 - 580 - 590 - 600 - 610 - 620 - 630 - 640 - 650
660 - 670 - 680 - 690 - 700 - 710 - 720 - 730 - 740 - 750 - 760 - 770 - 780
790 - 800 - 810 - 820 - 830 - 840 - 850 - 860 - 870 - 880 - 890 - 900 - 910
920 - 930 - 940 - 950 - 960 - 970 - 980 - 990 - 1000

Notes: ____

____

____

____

Reg Reps: 10 - 20 - 30 - 40 - 50 - 60 - 70 - 80 - 90 - 100 - 110 - 120 - 130
140 - 150 - 160 - 170 - 180 - 190 - 200 - 210 - 220 - 230 - 240 - 250 - 260
270 - 280 - 290 - 300 - 310 - 320 - 330 - 340 - 350 - 360 - 370 - 380 - 390
400 - 410 - 420 - 430 - 440 - 450 - 460 - 470 - 480 - 490 - 500 - 510 - 520
530 - 540 - 550 - 560 - 570 - 580 - 590 - 600 - 610 - 620 - 630 - 640 - 650
660 - 670 - 680 - 690 - 700 - 710 - 720 - 730 - 740 - 750 - 760 - 770 - 780
790 - 800 - 810 - 820 - 830 - 840 - 850 - 860 - 870 - 880 - 890 - 900 - 910
920 - 930 - 940 - 950 - 960 - 970 - 980 - 990 - 1000

Notes: ____

____

____

____

Reg Reps: 10 - 20 - 30 - 40 - 50 - 60 - 70 - 80 - 90 - 100 - 110 - 120 - 130
140 - 150 - 160 - 170 - 180 - 190 - 200 - 210 - 220 - 230 - 240 - 250 - 260
270 - 280 - 290 - 300 - 310 - 320 - 330 - 340 - 350 - 360 - 370 - 380 - 390
400 - 410 - 420 - 430 - 440 - 450 - 460 - 470 - 480 - 490 - 500 - 510 - 520
530 - 540 - 550 - 560 - 570 - 580 - 590 - 600 - 610 - 620 - 630 - 640 - 650
660 - 670 - 680 - 690 - 700 - 710 - 720 - 730 - 740 - 750 - 760 - 770 - 780
790 - 800 - 810 - 820 - 830 - 840 - 850 - 860 - 870 - 880 - 890 - 900 - 910
920 - 930 - 940 - 950 - 960 - 970 - 980 - 990 - 1000

Notes: ____

____

____

______________________________ **End Date:** ____________

Defense Against:__________________________ **Start Date:** __________

Notes:____

Slow Reps: 10 - 20 - 30 - 40 - 50 - 60 - 70 - 80 - 90 - 100 - 110 - 120 - 130
140 - 150 - 160 - 170 - 180 - 190 - 200 - 210 - 220 - 230 - 240 - 250 - 260
270 - 280 - 290 - 300 - 310 - 320 - 330 - 340 - 350 - 360 - 370 - 380 - 390
400 - 410 - 420 - 430 - 440 - 450 - 460 - 470 - 480 - 490 - 500 - 510 - 520
530 - 540 - 550 - 560 - 570 - 580 - 590 - 600 - 610 - 620 - 630 - 640 - 650
660 - 670 - 680 - 690 - 700 - 710 - 720 - 730 - 740 - 750 - 760 - 770 - 780
790 - 800 - 810 - 820 - 830 - 840 - 850 - 860 - 870 - 880 - 890 - 900 - 910
920 - 930 - 940 - 950 - 960 - 970 - 980 - 990 - 1000

Reg Reps: 10 - 20 - 30 - 40 - 50 - 60 - 70 - 80 - 90 - 100 - 110 - 120 - 130
140 - 150 - 160 - 170 - 180 - 190 - 200 - 210 - 220 - 230 - 240 - 250 - 260
270 - 280 - 290 - 300 - 310 - 320 - 330 - 340 - 350 - 360 - 370 - 380 - 390
400 - 410 - 420 - 430 - 440 - 450 - 460 - 470 - 480 - 490 - 500 - 510 - 520
530 - 540 - 550 - 560 - 570 - 580 - 590 - 600 - 610 - 620 - 630 - 640 - 650
660 - 670 - 680 - 690 - 700 - 710 - 720 - 730 - 740 - 750 - 760 - 770 - 780
790 - 800 - 810 - 820 - 830 - 840 - 850 - 860 - 870 - 880 - 890 - 900 - 910
920 - 930 - 940 - 950 - 960 - 970 - 980 - 990 - 1000

Notes: ____

Reg Reps: 10 - 20 - 30 - 40 - 50 - 60 - 70 - 80 - 90 - 100 - 110 - 120 - 130
140 - 150 - 160 - 170 - 180 - 190 - 200 - 210 - 220 - 230 - 240 - 250 - 260
270 - 280 - 290 - 300 - 310 - 320 - 330 - 340 - 350 - 360 - 370 - 380 - 390
400 - 410 - 420 - 430 - 440 - 450 - 460 - 470 - 480 - 490 - 500 - 510 - 520
530 - 540 - 550 - 560 - 570 - 580 - 590 - 600 - 610 - 620 - 630 - 640 - 650
660 - 670 - 680 - 690 - 700 - 710 - 720 - 730 - 740 - 750 - 760 - 770 - 780
790 - 800 - 810 - 820 - 830 - 840 - 850 - 860 - 870 - 880 - 890 - 900 - 910
920 - 930 - 940 - 950 - 960 - 970 - 980 - 990 - 1000

Notes: ____

Reg Reps: 10 - 20 - 30 - 40 - 50 - 60 - 70 - 80 - 90 - 100 - 110 - 120 - 130
140 - 150 - 160 - 170 - 180 - 190 - 200 - 210 - 220 - 230 - 240 - 250 - 260
270 - 280 - 290 - 300 - 310 - 320 - 330 - 340 - 350 - 360 - 370 - 380 - 390
400 - 410 - 420 - 430 - 440 - 450 - 460 - 470 - 480 - 490 - 500 - 510 - 520
530 - 540 - 550 - 560 - 570 - 580 - 590 - 600 - 610 - 620 - 630 - 640 - 650
660 - 670 - 680 - 690 - 700 - 710 - 720 - 730 - 740 - 750 - 760 - 770 - 780
790 - 800 - 810 - 820 - 830 - 840 - 850 - 860 - 870 - 880 - 890 - 900 - 910
920 - 930 - 940 - 950 - 960 - 970 - 980 - 990 - 1000

Notes: ______________________________

____

____

____

Reg Reps: 10 - 20 - 30 - 40 - 50 - 60 - 70 - 80 - 90 - 100 - 110 - 120 - 130
140 - 150 - 160 - 170 - 180 - 190 - 200 - 210 - 220 - 230 - 240 - 250 - 260
270 - 280 - 290 - 300 - 310 - 320 - 330 - 340 - 350 - 360 - 370 - 380 - 390
400 - 410 - 420 - 430 - 440 - 450 - 460 - 470 - 480 - 490 - 500 - 510 - 520
530 - 540 - 550 - 560 - 570 - 580 - 590 - 600 - 610 - 620 - 630 - 640 - 650
660 - 670 - 680 - 690 - 700 - 710 - 720 - 730 - 740 - 750 - 760 - 770 - 780
790 - 800 - 810 - 820 - 830 - 840 - 850 - 860 - 870 - 880 - 890 - 900 - 910
920 - 930 - 940 - 950 - 960 - 970 - 980 - 990 - 1000

Notes: ______________________________

____

____

____

Reg Reps: 10 - 20 - 30 - 40 - 50 - 60 - 70 - 80 - 90 - 100 - 110 - 120 - 130
140 - 150 - 160 - 170 - 180 - 190 - 200 - 210 - 220 - 230 - 240 - 250 - 260
270 - 280 - 290 - 300 - 310 - 320 - 330 - 340 - 350 - 360 - 370 - 380 - 390
400 - 410 - 420 - 430 - 440 - 450 - 460 - 470 - 480 - 490 - 500 - 510 - 520
530 - 540 - 550 - 560 - 570 - 580 - 590 - 600 - 610 - 620 - 630 - 640 - 650
660 - 670 - 680 - 690 - 700 - 710 - 720 - 730 - 740 - 750 - 760 - 770 - 780
790 - 800 - 810 - 820 - 830 - 840 - 850 - 860 - 870 - 880 - 890 - 900 - 910
920 - 930 - 940 - 950 - 960 - 970 - 980 - 990 - 1000

Notes: ______________________________

____

____

______________________________ **End Date:** ____________

Defense Against:_______________________ **Start Date:** __________

Notes:____

____

____

____

Slow Reps: 10 - 20 - 30 - 40 - 50 - 60 - 70 - 80 - 90 - 100 - 110 - 120 - 130
140 - 150 - 160 - 170 - 180 - 190 - 200 - 210 - 220 - 230 - 240 - 250 - 260
270 - 280 - 290 - 300 - 310 - 320 - 330 - 340 - 350 - 360 - 370 - 380 - 390
400 - 410 - 420 - 430 - 440 - 450 - 460 - 470 - 480 - 490 - 500 - 510 - 520
530 - 540 - 550 - 560 - 570 - 580 - 590 - 600 - 610 - 620 - 630 - 640 - 650
660 - 670 - 680 - 690 - 700 - 710 - 720 - 730 - 740 - 750 - 760 - 770 - 780
790 - 800 - 810 - 820 - 830 - 840 - 850 - 860 - 870 - 880 - 890 - 900 - 910
920 - 930 - 940 - 950 - 960 - 970 - 980 - 990 - 1000

Reg Reps: 10 - 20 - 30 - 40 - 50 - 60 - 70 - 80 - 90 - 100 - 110 - 120 - 130
140 - 150 - 160 - 170 - 180 - 190 - 200 - 210 - 220 - 230 - 240 - 250 - 260
270 - 280 - 290 - 300 - 310 - 320 - 330 - 340 - 350 - 360 - 370 - 380 - 390
400 - 410 - 420 - 430 - 440 - 450 - 460 - 470 - 480 - 490 - 500 - 510 - 520
530 - 540 - 550 - 560 - 570 - 580 - 590 - 600 - 610 - 620 - 630 - 640 - 650
660 - 670 - 680 - 690 - 700 - 710 - 720 - 730 - 740 - 750 - 760 - 770 - 780
790 - 800 - 810 - 820 - 830 - 840 - 850 - 860 - 870 - 880 - 890 - 900 - 910
920 - 930 - 940 - 950 - 960 - 970 - 980 - 990 - 1000

Notes: ____

____

____

____

Reg Reps: 10 - 20 - 30 - 40 - 50 - 60 - 70 - 80 - 90 - 100 - 110 - 120 - 130
140 - 150 - 160 - 170 - 180 - 190 - 200 - 210 - 220 - 230 - 240 - 250 - 260
270 - 280 - 290 - 300 - 310 - 320 - 330 - 340 - 350 - 360 - 370 - 380 - 390
400 - 410 - 420 - 430 - 440 - 450 - 460 - 470 - 480 - 490 - 500 - 510 - 520
530 - 540 - 550 - 560 - 570 - 580 - 590 - 600 - 610 - 620 - 630 - 640 - 650
660 - 670 - 680 - 690 - 700 - 710 - 720 - 730 - 740 - 750 - 760 - 770 - 780
790 - 800 - 810 - 820 - 830 - 840 - 850 - 860 - 870 - 880 - 890 - 900 - 910
920 - 930 - 940 - 950 - 960 - 970 - 980 - 990 - 1000

Notes: ____

____

____

____

Reg Reps: 10 - 20 - 30 - 40 - 50 - 60 - 70 - 80 - 90 - 100 - 110 - 120 - 130
140 - 150 - 160 - 170 - 180 - 190 - 200 - 210 - 220 - 230 - 240 - 250 - 260
270 - 280 - 290 - 300 - 310 - 320 - 330 - 340 - 350 - 360 - 370 - 380 - 390
400 - 410 - 420 - 430 - 440 - 450 - 460 - 470 - 480 - 490 - 500 - 510 - 520
530 - 540 - 550 - 560 - 570 - 580 - 590 - 600 - 610 - 620 - 630 - 640 - 650
660 - 670 - 680 - 690 - 700 - 710 - 720 - 730 - 740 - 750 - 760 - 770 - 780
790 - 800 - 810 - 820 - 830 - 840 - 850 - 860 - 870 - 880 - 890 - 900 - 910
920 - 930 - 940 - 950 - 960 - 970 - 980 - 990 - 1000

Notes: ______________________________

____

____

____

Reg Reps: 10 - 20 - 30 - 40 - 50 - 60 - 70 - 80 - 90 - 100 - 110 - 120 - 130
140 - 150 - 160 - 170 - 180 - 190 - 200 - 210 - 220 - 230 - 240 - 250 - 260
270 - 280 - 290 - 300 - 310 - 320 - 330 - 340 - 350 - 360 - 370 - 380 - 390
400 - 410 - 420 - 430 - 440 - 450 - 460 - 470 - 480 - 490 - 500 - 510 - 520
530 - 540 - 550 - 560 - 570 - 580 - 590 - 600 - 610 - 620 - 630 - 640 - 650
660 - 670 - 680 - 690 - 700 - 710 - 720 - 730 - 740 - 750 - 760 - 770 - 780
790 - 800 - 810 - 820 - 830 - 840 - 850 - 860 - 870 - 880 - 890 - 900 - 910
920 - 930 - 940 - 950 - 960 - 970 - 980 - 990 - 1000

Notes: ______________________________

____

____

____

Reg Reps: 10 - 20 - 30 - 40 - 50 - 60 - 70 - 80 - 90 - 100 - 110 - 120 - 130
140 - 150 - 160 - 170 - 180 - 190 - 200 - 210 - 220 - 230 - 240 - 250 - 260
270 - 280 - 290 - 300 - 310 - 320 - 330 - 340 - 350 - 360 - 370 - 380 - 390
400 - 410 - 420 - 430 - 440 - 450 - 460 - 470 - 480 - 490 - 500 - 510 - 520
530 - 540 - 550 - 560 - 570 - 580 - 590 - 600 - 610 - 620 - 630 - 640 - 650
660 - 670 - 680 - 690 - 700 - 710 - 720 - 730 - 740 - 750 - 760 - 770 - 780
790 - 800 - 810 - 820 - 830 - 840 - 850 - 860 - 870 - 880 - 890 - 900 - 910
920 - 930 - 940 - 950 - 960 - 970 - 980 - 990 - 1000

Notes: ______________________________

____

____

______________________________ **End Date:** ____________

Defense Against:____________________ **Start Date:** __________
Notes:____

Slow Reps: 10 - 20 - 30 - 40 - 50 - 60 - 70 - 80 - 90 - 100 - 110 - 120 - 130
140 - 150 - 160 - 170 - 180 - 190 - 200 - 210 - 220 - 230 - 240 - 250 - 260
270 - 280 - 290 - 300 - 310 - 320 - 330 - 340 - 350 - 360 - 370 - 380 - 390
400 - 410 - 420 - 430 - 440 - 450 - 460 - 470 - 480 - 490 - 500 - 510 - 520
530 - 540 - 550 - 560 - 570 - 580 - 590 - 600 - 610 - 620 - 630 - 640 - 650
660 - 670 - 680 - 690 - 700 - 710 - 720 - 730 - 740 - 750 - 760 - 770 - 780
790 - 800 - 810 - 820 - 830 - 840 - 850 - 860 - 870 - 880 - 890 - 900 - 910
920 - 930 - 940 - 950 - 960 - 970 - 980 - 990 - 1000

Reg Reps: 10 - 20 - 30 - 40 - 50 - 60 - 70 - 80 - 90 - 100 - 110 - 120 - 130
140 - 150 - 160 - 170 - 180 - 190 - 200 - 210 - 220 - 230 - 240 - 250 - 260
270 - 280 - 290 - 300 - 310 - 320 - 330 - 340 - 350 - 360 - 370 - 380 - 390
400 - 410 - 420 - 430 - 440 - 450 - 460 - 470 - 480 - 490 - 500 - 510 - 520
530 - 540 - 550 - 560 - 570 - 580 - 590 - 600 - 610 - 620 - 630 - 640 - 650
660 - 670 - 680 - 690 - 700 - 710 - 720 - 730 - 740 - 750 - 760 - 770 - 780
790 - 800 - 810 - 820 - 830 - 840 - 850 - 860 - 870 - 880 - 890 - 900 - 910
920 - 930 - 940 - 950 - 960 - 970 - 980 - 990 - 1000
Notes: ____

Reg Reps: 10 - 20 - 30 - 40 - 50 - 60 - 70 - 80 - 90 - 100 - 110 - 120 - 130
140 - 150 - 160 - 170 - 180 - 190 - 200 - 210 - 220 - 230 - 240 - 250 - 260
270 - 280 - 290 - 300 - 310 - 320 - 330 - 340 - 350 - 360 - 370 - 380 - 390
400 - 410 - 420 - 430 - 440 - 450 - 460 - 470 - 480 - 490 - 500 - 510 - 520
530 - 540 - 550 - 560 - 570 - 580 - 590 - 600 - 610 - 620 - 630 - 640 - 650
660 - 670 - 680 - 690 - 700 - 710 - 720 - 730 - 740 - 750 - 760 - 770 - 780
790 - 800 - 810 - 820 - 830 - 840 - 850 - 860 - 870 - 880 - 890 - 900 - 910
920 - 930 - 940 - 950 - 960 - 970 - 980 - 990 - 1000
Notes: ____

Reg Reps: 10 - 20 - 30 - 40 - 50 - 60 - 70 - 80 - 90 - 100 - 110 - 120 - 130
140 - 150 - 160 - 170 - 180 - 190 - 200 - 210 - 220 - 230 - 240 - 250 - 260
270 - 280 - 290 - 300 - 310 - 320 - 330 - 340 - 350 - 360 - 370 - 380 - 390
400 - 410 - 420 - 430 - 440 - 450 - 460 - 470 - 480 - 490 - 500 - 510 - 520
530 - 540 - 550 - 560 - 570 - 580 - 590 - 600 - 610 - 620 - 630 - 640 - 650
660 - 670 - 680 - 690 - 700 - 710 - 720 - 730 - 740 - 750 - 760 - 770 - 780
790 - 800 - 810 - 820 - 830 - 840 - 850 - 860 - 870 - 880 - 890 - 900 - 910
920 - 930 - 940 - 950 - 960 - 970 - 980 - 990 - 1000

Notes: ____

____

____

____

Reg Reps: 10 - 20 - 30 - 40 - 50 - 60 - 70 - 80 - 90 - 100 - 110 - 120 - 130
140 - 150 - 160 - 170 - 180 - 190 - 200 - 210 - 220 - 230 - 240 - 250 - 260
270 - 280 - 290 - 300 - 310 - 320 - 330 - 340 - 350 - 360 - 370 - 380 - 390
400 - 410 - 420 - 430 - 440 - 450 - 460 - 470 - 480 - 490 - 500 - 510 - 520
530 - 540 - 550 - 560 - 570 - 580 - 590 - 600 - 610 - 620 - 630 - 640 - 650
660 - 670 - 680 - 690 - 700 - 710 - 720 - 730 - 740 - 750 - 760 - 770 - 780
790 - 800 - 810 - 820 - 830 - 840 - 850 - 860 - 870 - 880 - 890 - 900 - 910
920 - 930 - 940 - 950 - 960 - 970 - 980 - 990 - 1000

Notes: ____

____

____

____

Reg Reps: 10 - 20 - 30 - 40 - 50 - 60 - 70 - 80 - 90 - 100 - 110 - 120 - 130
140 - 150 - 160 - 170 - 180 - 190 - 200 - 210 - 220 - 230 - 240 - 250 - 260
270 - 280 - 290 - 300 - 310 - 320 - 330 - 340 - 350 - 360 - 370 - 380 - 390
400 - 410 - 420 - 430 - 440 - 450 - 460 - 470 - 480 - 490 - 500 - 510 - 520
530 - 540 - 550 - 560 - 570 - 580 - 590 - 600 - 610 - 620 - 630 - 640 - 650
660 - 670 - 680 - 690 - 700 - 710 - 720 - 730 - 740 - 750 - 760 - 770 - 780
790 - 800 - 810 - 820 - 830 - 840 - 850 - 860 - 870 - 880 - 890 - 900 - 910
920 - 930 - 940 - 950 - 960 - 970 - 980 - 990 - 1000

Notes: ____

____

____

______________________________ **End Date:** ____________

Defense Against:__________________________ **Start Date:** __________

Notes:____

Slow Reps: 10 - 20 - 30 - 40 - 50 - 60 - 70 - 80 - 90 - 100 - 110 - 120 - 130
140 - 150 - 160 - 170 - 180 - 190 - 200 - 210 - 220 - 230 - 240 - 250 - 260
270 - 280 - 290 - 300 - 310 - 320 - 330 - 340 - 350 - 360 - 370 - 380 - 390
400 - 410 - 420 - 430 - 440 - 450 - 460 - 470 - 480 - 490 - 500 - 510 - 520
530 - 540 - 550 - 560 - 570 - 580 - 590 - 600 - 610 - 620 - 630 - 640 - 650
660 - 670 - 680 - 690 - 700 - 710 - 720 - 730 - 740 - 750 - 760 - 770 - 780
790 - 800 - 810 - 820 - 830 - 840 - 850 - 860 - 870 - 880 - 890 - 900 - 910
920 - 930 - 940 - 950 - 960 - 970 - 980 - 990 - 1000

Reg Reps: 10 - 20 - 30 - 40 - 50 - 60 - 70 - 80 - 90 - 100 - 110 - 120 - 130
140 - 150 - 160 - 170 - 180 - 190 - 200 - 210 - 220 - 230 - 240 - 250 - 260
270 - 280 - 290 - 300 - 310 - 320 - 330 - 340 - 350 - 360 - 370 - 380 - 390
400 - 410 - 420 - 430 - 440 - 450 - 460 - 470 - 480 - 490 - 500 - 510 - 520
530 - 540 - 550 - 560 - 570 - 580 - 590 - 600 - 610 - 620 - 630 - 640 - 650
660 - 670 - 680 - 690 - 700 - 710 - 720 - 730 - 740 - 750 - 760 - 770 - 780
790 - 800 - 810 - 820 - 830 - 840 - 850 - 860 - 870 - 880 - 890 - 900 - 910
920 - 930 - 940 - 950 - 960 - 970 - 980 - 990 - 1000

Notes: ____

Reg Reps: 10 - 20 - 30 - 40 - 50 - 60 - 70 - 80 - 90 - 100 - 110 - 120 - 130
140 - 150 - 160 - 170 - 180 - 190 - 200 - 210 - 220 - 230 - 240 - 250 - 260
270 - 280 - 290 - 300 - 310 - 320 - 330 - 340 - 350 - 360 - 370 - 380 - 390
400 - 410 - 420 - 430 - 440 - 450 - 460 - 470 - 480 - 490 - 500 - 510 - 520
530 - 540 - 550 - 560 - 570 - 580 - 590 - 600 - 610 - 620 - 630 - 640 - 650
660 - 670 - 680 - 690 - 700 - 710 - 720 - 730 - 740 - 750 - 760 - 770 - 780
790 - 800 - 810 - 820 - 830 - 840 - 850 - 860 - 870 - 880 - 890 - 900 - 910
920 - 930 - 940 - 950 - 960 - 970 - 980 - 990 - 1000

Notes: ____

Reg Reps: 10 - 20 - 30 - 40 - 50 - 60 - 70 - 80 - 90 - 100 - 110 - 120 - 130
140 - 150 - 160 - 170 - 180 - 190 - 200 - 210 - 220 - 230 - 240 - 250 - 260
270 - 280 - 290 - 300 - 310 - 320 - 330 - 340 - 350 - 360 - 370 - 380 - 390
400 - 410 - 420 - 430 - 440 - 450 - 460 - 470 - 480 - 490 - 500 - 510 - 520
530 - 540 - 550 - 560 - 570 - 580 - 590 - 600 - 610 - 620 - 630 - 640 - 650
660 - 670 - 680 - 690 - 700 - 710 - 720 - 730 - 740 - 750 - 760 - 770 - 780
790 - 800 - 810 - 820 - 830 - 840 - 850 - 860 - 870 - 880 - 890 - 900 - 910
920 - 930 - 940 - 950 - 960 - 970 - 980 - 990 - 1000

Notes: ______________________________

____

____

____

Reg Reps: 10 - 20 - 30 - 40 - 50 - 60 - 70 - 80 - 90 - 100 - 110 - 120 - 130
140 - 150 - 160 - 170 - 180 - 190 - 200 - 210 - 220 - 230 - 240 - 250 - 260
270 - 280 - 290 - 300 - 310 - 320 - 330 - 340 - 350 - 360 - 370 - 380 - 390
400 - 410 - 420 - 430 - 440 - 450 - 460 - 470 - 480 - 490 - 500 - 510 - 520
530 - 540 - 550 - 560 - 570 - 580 - 590 - 600 - 610 - 620 - 630 - 640 - 650
660 - 670 - 680 - 690 - 700 - 710 - 720 - 730 - 740 - 750 - 760 - 770 - 780
790 - 800 - 810 - 820 - 830 - 840 - 850 - 860 - 870 - 880 - 890 - 900 - 910
920 - 930 - 940 - 950 - 960 - 970 - 980 - 990 - 1000

Notes: ______________________________

____

____

____

Reg Reps: 10 - 20 - 30 - 40 - 50 - 60 - 70 - 80 - 90 - 100 - 110 - 120 - 130
140 - 150 - 160 - 170 - 180 - 190 - 200 - 210 - 220 - 230 - 240 - 250 - 260
270 - 280 - 290 - 300 - 310 - 320 - 330 - 340 - 350 - 360 - 370 - 380 - 390
400 - 410 - 420 - 430 - 440 - 450 - 460 - 470 - 480 - 490 - 500 - 510 - 520
530 - 540 - 550 - 560 - 570 - 580 - 590 - 600 - 610 - 620 - 630 - 640 - 650
660 - 670 - 680 - 690 - 700 - 710 - 720 - 730 - 740 - 750 - 760 - 770 - 780
790 - 800 - 810 - 820 - 830 - 840 - 850 - 860 - 870 - 880 - 890 - 900 - 910
920 - 930 - 940 - 950 - 960 - 970 - 980 - 990 - 1000

Notes: ______________________________

____

____

______________________________ **End Date:** ____________

Defense Against:__________________________ **Start Date:** __________

Notes:____

Slow Reps: 10 - 20 - 30 - 40 - 50 - 60 - 70 - 80 - 90 - 100 - 110 - 120 - 130
140 - 150 - 160 - 170 - 180 - 190 - 200 - 210 - 220 - 230 - 240 - 250 - 260
270 - 280 - 290 - 300 - 310 - 320 - 330 - 340 - 350 - 360 - 370 - 380 - 390
400 - 410 - 420 - 430 - 440 - 450 - 460 - 470 - 480 - 490 - 500 - 510 - 520
530 - 540 - 550 - 560 - 570 - 580 - 590 - 600 - 610 - 620 - 630 - 640 - 650
660 - 670 - 680 - 690 - 700 - 710 - 720 - 730 - 740 - 750 - 760 - 770 - 780
790 - 800 - 810 - 820 - 830 - 840 - 850 - 860 - 870 - 880 - 890 - 900 - 910
920 - 930 - 940 - 950 - 960 - 970 - 980 - 990 - 1000

Reg Reps: 10 - 20 - 30 - 40 - 50 - 60 - 70 - 80 - 90 - 100 - 110 - 120 - 130
140 - 150 - 160 - 170 - 180 - 190 - 200 - 210 - 220 - 230 - 240 - 250 - 260
270 - 280 - 290 - 300 - 310 - 320 - 330 - 340 - 350 - 360 - 370 - 380 - 390
400 - 410 - 420 - 430 - 440 - 450 - 460 - 470 - 480 - 490 - 500 - 510 - 520
530 - 540 - 550 - 560 - 570 - 580 - 590 - 600 - 610 - 620 - 630 - 640 - 650
660 - 670 - 680 - 690 - 700 - 710 - 720 - 730 - 740 - 750 - 760 - 770 - 780
790 - 800 - 810 - 820 - 830 - 840 - 850 - 860 - 870 - 880 - 890 - 900 - 910
920 - 930 - 940 - 950 - 960 - 970 - 980 - 990 - 1000

Notes: ____

Reg Reps: 10 - 20 - 30 - 40 - 50 - 60 - 70 - 80 - 90 - 100 - 110 - 120 - 130
140 - 150 - 160 - 170 - 180 - 190 - 200 - 210 - 220 - 230 - 240 - 250 - 260
270 - 280 - 290 - 300 - 310 - 320 - 330 - 340 - 350 - 360 - 370 - 380 - 390
400 - 410 - 420 - 430 - 440 - 450 - 460 - 470 - 480 - 490 - 500 - 510 - 520
530 - 540 - 550 - 560 - 570 - 580 - 590 - 600 - 610 - 620 - 630 - 640 - 650
660 - 670 - 680 - 690 - 700 - 710 - 720 - 730 - 740 - 750 - 760 - 770 - 780
790 - 800 - 810 - 820 - 830 - 840 - 850 - 860 - 870 - 880 - 890 - 900 - 910
920 - 930 - 940 - 950 - 960 - 970 - 980 - 990 - 1000

Notes: ____

Reg Reps: 10 - 20 - 30 - 40 - 50 - 60 - 70 - 80 - 90 - 100 - 110 - 120 - 130
140 - 150 - 160 - 170 - 180 - 190 - 200 - 210 - 220 - 230 - 240 - 250 - 260
270 - 280 - 290 - 300 - 310 - 320 - 330 - 340 - 350 - 360 - 370 - 380 - 390
400 - 410 - 420 - 430 - 440 - 450 - 460 - 470 - 480 - 490 - 500 - 510 - 520
530 - 540 - 550 - 560 - 570 - 580 - 590 - 600 - 610 - 620 - 630 - 640 - 650
660 - 670 - 680 - 690 - 700 - 710 - 720 - 730 - 740 - 750 - 760 - 770 - 780
790 - 800 - 810 - 820 - 830 - 840 - 850 - 860 - 870 - 880 - 890 - 900 - 910
920 - 930 - 940 - 950 - 960 - 970 - 980 - 990 - 1000

Notes: ____

____

____

____

Reg Reps: 10 - 20 - 30 - 40 - 50 - 60 - 70 - 80 - 90 - 100 - 110 - 120 - 130
140 - 150 - 160 - 170 - 180 - 190 - 200 - 210 - 220 - 230 - 240 - 250 - 260
270 - 280 - 290 - 300 - 310 - 320 - 330 - 340 - 350 - 360 - 370 - 380 - 390
400 - 410 - 420 - 430 - 440 - 450 - 460 - 470 - 480 - 490 - 500 - 510 - 520
530 - 540 - 550 - 560 - 570 - 580 - 590 - 600 - 610 - 620 - 630 - 640 - 650
660 - 670 - 680 - 690 - 700 - 710 - 720 - 730 - 740 - 750 - 760 - 770 - 780
790 - 800 - 810 - 820 - 830 - 840 - 850 - 860 - 870 - 880 - 890 - 900 - 910
920 - 930 - 940 - 950 - 960 - 970 - 980 - 990 - 1000

Notes: ____

____

____

____

Reg Reps: 10 - 20 - 30 - 40 - 50 - 60 - 70 - 80 - 90 - 100 - 110 - 120 - 130
140 - 150 - 160 - 170 - 180 - 190 - 200 - 210 - 220 - 230 - 240 - 250 - 260
270 - 280 - 290 - 300 - 310 - 320 - 330 - 340 - 350 - 360 - 370 - 380 - 390
400 - 410 - 420 - 430 - 440 - 450 - 460 - 470 - 480 - 490 - 500 - 510 - 520
530 - 540 - 550 - 560 - 570 - 580 - 590 - 600 - 610 - 620 - 630 - 640 - 650
660 - 670 - 680 - 690 - 700 - 710 - 720 - 730 - 740 - 750 - 760 - 770 - 780
790 - 800 - 810 - 820 - 830 - 840 - 850 - 860 - 870 - 880 - 890 - 900 - 910
920 - 930 - 940 - 950 - 960 - 970 - 980 - 990 - 1000

Notes: ____

____

____

______________________________ **End Date:** ____________

Defense Against:____________________________ **Start Date:** ___________

Notes:____

Slow Reps: 10 - 20 - 30 - 40 - 50 - 60 - 70 - 80 - 90 - 100 - 110 - 120 - 130
140 - 150 - 160 - 170 - 180 - 190 - 200 - 210 - 220 - 230 - 240 - 250 - 260
270 - 280 - 290 - 300 - 310 - 320 - 330 - 340 - 350 - 360 - 370 - 380 - 390
400 - 410 - 420 - 430 - 440 - 450 - 460 - 470 - 480 - 490 - 500 - 510 - 520
530 - 540 - 550 - 560 - 570 - 580 - 590 - 600 - 610 - 620 - 630 - 640 - 650
660 - 670 - 680 - 690 - 700 - 710 - 720 - 730 - 740 - 750 - 760 - 770 - 780
790 - 800 - 810 - 820 - 830 - 840 - 850 - 860 - 870 - 880 - 890 - 900 - 910
920 - 930 - 940 - 950 - 960 - 970 - 980 - 990 - 1000

Reg Reps: 10 - 20 - 30 - 40 - 50 - 60 - 70 - 80 - 90 - 100 - 110 - 120 - 130
140 - 150 - 160 - 170 - 180 - 190 - 200 - 210 - 220 - 230 - 240 - 250 - 260
270 - 280 - 290 - 300 - 310 - 320 - 330 - 340 - 350 - 360 - 370 - 380 - 390
400 - 410 - 420 - 430 - 440 - 450 - 460 - 470 - 480 - 490 - 500 - 510 - 520
530 - 540 - 550 - 560 - 570 - 580 - 590 - 600 - 610 - 620 - 630 - 640 - 650
660 - 670 - 680 - 690 - 700 - 710 - 720 - 730 - 740 - 750 - 760 - 770 - 780
790 - 800 - 810 - 820 - 830 - 840 - 850 - 860 - 870 - 880 - 890 - 900 - 910
920 - 930 - 940 - 950 - 960 - 970 - 980 - 990 - 1000

Notes: ____

Reg Reps: 10 - 20 - 30 - 40 - 50 - 60 - 70 - 80 - 90 - 100 - 110 - 120 - 130
140 - 150 - 160 - 170 - 180 - 190 - 200 - 210 - 220 - 230 - 240 - 250 - 260
270 - 280 - 290 - 300 - 310 - 320 - 330 - 340 - 350 - 360 - 370 - 380 - 390
400 - 410 - 420 - 430 - 440 - 450 - 460 - 470 - 480 - 490 - 500 - 510 - 520
530 - 540 - 550 - 560 - 570 - 580 - 590 - 600 - 610 - 620 - 630 - 640 - 650
660 - 670 - 680 - 690 - 700 - 710 - 720 - 730 - 740 - 750 - 760 - 770 - 780
790 - 800 - 810 - 820 - 830 - 840 - 850 - 860 - 870 - 880 - 890 - 900 - 910
920 - 930 - 940 - 950 - 960 - 970 - 980 - 990 - 1000

Notes: ____

Reg Reps: 10 - 20 - 30 - 40 - 50 - 60 - 70 - 80 - 90 - 100 - 110 - 120 - 130
140 - 150 - 160 - 170 - 180 - 190 - 200 - 210 - 220 - 230 - 240 - 250 - 260
270 - 280 - 290 - 300 - 310 - 320 - 330 - 340 - 350 - 360 - 370 - 380 - 390
400 - 410 - 420 - 430 - 440 - 450 - 460 - 470 - 480 - 490 - 500 - 510 - 520
530 - 540 - 550 - 560 - 570 - 580 - 590 - 600 - 610 - 620 - 630 - 640 - 650
660 - 670 - 680 - 690 - 700 - 710 - 720 - 730 - 740 - 750 - 760 - 770 - 780
790 - 800 - 810 - 820 - 830 - 840 - 850 - 860 - 870 - 880 - 890 - 900 - 910
920 - 930 - 940 - 950 - 960 - 970 - 980 - 990 - 1000

Notes: ______________________________

____

____

____

Reg Reps: 10 - 20 - 30 - 40 - 50 - 60 - 70 - 80 - 90 - 100 - 110 - 120 - 130
140 - 150 - 160 - 170 - 180 - 190 - 200 - 210 - 220 - 230 - 240 - 250 - 260
270 - 280 - 290 - 300 - 310 - 320 - 330 - 340 - 350 - 360 - 370 - 380 - 390
400 - 410 - 420 - 430 - 440 - 450 - 460 - 470 - 480 - 490 - 500 - 510 - 520
530 - 540 - 550 - 560 - 570 - 580 - 590 - 600 - 610 - 620 - 630 - 640 - 650
660 - 670 - 680 - 690 - 700 - 710 - 720 - 730 - 740 - 750 - 760 - 770 - 780
790 - 800 - 810 - 820 - 830 - 840 - 850 - 860 - 870 - 880 - 890 - 900 - 910
920 - 930 - 940 - 950 - 960 - 970 - 980 - 990 - 1000

Notes: ______________________________

____

____

____

Reg Reps: 10 - 20 - 30 - 40 - 50 - 60 - 70 - 80 - 90 - 100 - 110 - 120 - 130
140 - 150 - 160 - 170 - 180 - 190 - 200 - 210 - 220 - 230 - 240 - 250 - 260
270 - 280 - 290 - 300 - 310 - 320 - 330 - 340 - 350 - 360 - 370 - 380 - 390
400 - 410 - 420 - 430 - 440 - 450 - 460 - 470 - 480 - 490 - 500 - 510 - 520
530 - 540 - 550 - 560 - 570 - 580 - 590 - 600 - 610 - 620 - 630 - 640 - 650
660 - 670 - 680 - 690 - 700 - 710 - 720 - 730 - 740 - 750 - 760 - 770 - 780
790 - 800 - 810 - 820 - 830 - 840 - 850 - 860 - 870 - 880 - 890 - 900 - 910
920 - 930 - 940 - 950 - 960 - 970 - 980 - 990 - 1000

Notes: ______________________________

____

____

______________________________ **End Date:** ____________

Defense Against:________________________ **Start Date:** __________
Notes:____

Slow Reps: 10 - 20 - 30 - 40 - 50 - 60 - 70 - 80 - 90 - 100 - 110 - 120 - 130
140 - 150 - 160 - 170 - 180 - 190 - 200 - 210 - 220 - 230 - 240 - 250 - 260
270 - 280 - 290 - 300 - 310 - 320 - 330 - 340 - 350 - 360 - 370 - 380 - 390
400 - 410 - 420 - 430 - 440 - 450 - 460 - 470 - 480 - 490 - 500 - 510 - 520
530 - 540 - 550 - 560 - 570 - 580 - 590 - 600 - 610 - 620 - 630 - 640 - 650
660 - 670 - 680 - 690 - 700 - 710 - 720 - 730 - 740 - 750 - 760 - 770 - 780
790 - 800 - 810 - 820 - 830 - 840 - 850 - 860 - 870 - 880 - 890 - 900 - 910
920 - 930 - 940 - 950 - 960 - 970 - 980 - 990 - 1000

Reg Reps: 10 - 20 - 30 - 40 - 50 - 60 - 70 - 80 - 90 - 100 - 110 - 120 - 130
140 - 150 - 160 - 170 - 180 - 190 - 200 - 210 - 220 - 230 - 240 - 250 - 260
270 - 280 - 290 - 300 - 310 - 320 - 330 - 340 - 350 - 360 - 370 - 380 - 390
400 - 410 - 420 - 430 - 440 - 450 - 460 - 470 - 480 - 490 - 500 - 510 - 520
530 - 540 - 550 - 560 - 570 - 580 - 590 - 600 - 610 - 620 - 630 - 640 - 650
660 - 670 - 680 - 690 - 700 - 710 - 720 - 730 - 740 - 750 - 760 - 770 - 780
790 - 800 - 810 - 820 - 830 - 840 - 850 - 860 - 870 - 880 - 890 - 900 - 910
920 - 930 - 940 - 950 - 960 - 970 - 980 - 990 - 1000
Notes: ____

Reg Reps: 10 - 20 - 30 - 40 - 50 - 60 - 70 - 80 - 90 - 100 - 110 - 120 - 130
140 - 150 - 160 - 170 - 180 - 190 - 200 - 210 - 220 - 230 - 240 - 250 - 260
270 - 280 - 290 - 300 - 310 - 320 - 330 - 340 - 350 - 360 - 370 - 380 - 390
400 - 410 - 420 - 430 - 440 - 450 - 460 - 470 - 480 - 490 - 500 - 510 - 520
530 - 540 - 550 - 560 - 570 - 580 - 590 - 600 - 610 - 620 - 630 - 640 - 650
660 - 670 - 680 - 690 - 700 - 710 - 720 - 730 - 740 - 750 - 760 - 770 - 780
790 - 800 - 810 - 820 - 830 - 840 - 850 - 860 - 870 - 880 - 890 - 900 - 910
920 - 930 - 940 - 950 - 960 - 970 - 980 - 990 - 1000
Notes: ____

Reg Reps: 10 - 20 - 30 - 40 - 50 - 60 - 70 - 80 - 90 - 100 - 110 - 120 - 130 140 - 150 - 160 - 170 - 180 - 190 - 200 - 210 - 220 - 230 - 240 - 250 - 260 270 - 280 - 290 - 300 - 310 - 320 - 330 - 340 - 350 - 360 - 370 - 380 - 390 400 - 410 - 420 - 430 - 440 - 450 - 460 - 470 - 480 - 490 - 500 - 510 - 520 530 - 540 - 550 - 560 - 570 - 580 - 590 - 600 - 610 - 620 - 630 - 640 - 650 660 - 670 - 680 - 690 - 700 - 710 - 720 - 730 - 740 - 750 - 760 - 770 - 780 790 - 800 - 810 - 820 - 830 - 840 - 850 - 860 - 870 - 880 - 890 - 900 - 910 920 - 930 - 940 - 950 - 960 - 970 - 980 - 990 - 1000

Notes: ______________________________

____

____

____

Reg Reps: 10 - 20 - 30 - 40 - 50 - 60 - 70 - 80 - 90 - 100 - 110 - 120 - 130 140 - 150 - 160 - 170 - 180 - 190 - 200 - 210 - 220 - 230 - 240 - 250 - 260 270 - 280 - 290 - 300 - 310 - 320 - 330 - 340 - 350 - 360 - 370 - 380 - 390 400 - 410 - 420 - 430 - 440 - 450 - 460 - 470 - 480 - 490 - 500 - 510 - 520 530 - 540 - 550 - 560 - 570 - 580 - 590 - 600 - 610 - 620 - 630 - 640 - 650 660 - 670 - 680 - 690 - 700 - 710 - 720 - 730 - 740 - 750 - 760 - 770 - 780 790 - 800 - 810 - 820 - 830 - 840 - 850 - 860 - 870 - 880 - 890 - 900 - 910 920 - 930 - 940 - 950 - 960 - 970 - 980 - 990 - 1000

Notes: ______________________________

____

____

____

Reg Reps: 10 - 20 - 30 - 40 - 50 - 60 - 70 - 80 - 90 - 100 - 110 - 120 - 130 140 - 150 - 160 - 170 - 180 - 190 - 200 - 210 - 220 - 230 - 240 - 250 - 260 270 - 280 - 290 - 300 - 310 - 320 - 330 - 340 - 350 - 360 - 370 - 380 - 390 400 - 410 - 420 - 430 - 440 - 450 - 460 - 470 - 480 - 490 - 500 - 510 - 520 530 - 540 - 550 - 560 - 570 - 580 - 590 - 600 - 610 - 620 - 630 - 640 - 650 660 - 670 - 680 - 690 - 700 - 710 - 720 - 730 - 740 - 750 - 760 - 770 - 780 790 - 800 - 810 - 820 - 830 - 840 - 850 - 860 - 870 - 880 - 890 - 900 - 910 920 - 930 - 940 - 950 - 960 - 970 - 980 - 990 - 1000

Notes: ______________________________

____

____

______________________________ **End Date:** ______________

Defense Against: ______________________________ **Start Date:** ___________

Notes: ____

____

____

____

Slow Reps: 10 - 20 - 30 - 40 - 50 - 60 - 70 - 80 - 90 - 100 - 110 - 120 - 130
140 - 150 - 160 - 170 - 180 - 190 - 200 - 210 - 220 - 230 - 240 - 250 - 260
270 - 280 - 290 - 300 - 310 - 320 - 330 - 340 - 350 - 360 - 370 - 380 - 390
400 - 410 - 420 - 430 - 440 - 450 - 460 - 470 - 480 - 490 - 500 - 510 - 520
530 - 540 - 550 - 560 - 570 - 580 - 590 - 600 - 610 - 620 - 630 - 640 - 650
660 - 670 - 680 - 690 - 700 - 710 - 720 - 730 - 740 - 750 - 760 - 770 - 780
790 - 800 - 810 - 820 - 830 - 840 - 850 - 860 - 870 - 880 - 890 - 900 - 910
920 - 930 - 940 - 950 - 960 - 970 - 980 - 990 - 1000

Reg Reps: 10 - 20 - 30 - 40 - 50 - 60 - 70 - 80 - 90 - 100 - 110 - 120 - 130
140 - 150 - 160 - 170 - 180 - 190 - 200 - 210 - 220 - 230 - 240 - 250 - 260
270 - 280 - 290 - 300 - 310 - 320 - 330 - 340 - 350 - 360 - 370 - 380 - 390
400 - 410 - 420 - 430 - 440 - 450 - 460 - 470 - 480 - 490 - 500 - 510 - 520
530 - 540 - 550 - 560 - 570 - 580 - 590 - 600 - 610 - 620 - 630 - 640 - 650
660 - 670 - 680 - 690 - 700 - 710 - 720 - 730 - 740 - 750 - 760 - 770 - 780
790 - 800 - 810 - 820 - 830 - 840 - 850 - 860 - 870 - 880 - 890 - 900 - 910
920 - 930 - 940 - 950 - 960 - 970 - 980 - 990 - 1000

Notes: ____

____

____

____

Reg Reps: 10 - 20 - 30 - 40 - 50 - 60 - 70 - 80 - 90 - 100 - 110 - 120 - 130
140 - 150 - 160 - 170 - 180 - 190 - 200 - 210 - 220 - 230 - 240 - 250 - 260
270 - 280 - 290 - 300 - 310 - 320 - 330 - 340 - 350 - 360 - 370 - 380 - 390
400 - 410 - 420 - 430 - 440 - 450 - 460 - 470 - 480 - 490 - 500 - 510 - 520
530 - 540 - 550 - 560 - 570 - 580 - 590 - 600 - 610 - 620 - 630 - 640 - 650
660 - 670 - 680 - 690 - 700 - 710 - 720 - 730 - 740 - 750 - 760 - 770 - 780
790 - 800 - 810 - 820 - 830 - 840 - 850 - 860 - 870 - 880 - 890 - 900 - 910
920 - 930 - 940 - 950 - 960 - 970 - 980 - 990 - 1000

Notes: ____

____

____

____

Reg Reps: 10 - 20 - 30 - 40 - 50 - 60 - 70 - 80 - 90 - 100 - 110 - 120 - 130
140 - 150 - 160 - 170 - 180 - 190 - 200 - 210 - 220 - 230 - 240 - 250 - 260
270 - 280 - 290 - 300 - 310 - 320 - 330 - 340 - 350 - 360 - 370 - 380 - 390
400 - 410 - 420 - 430 - 440 - 450 - 460 - 470 - 480 - 490 - 500 - 510 - 520
530 - 540 - 550 - 560 - 570 - 580 - 590 - 600 - 610 - 620 - 630 - 640 - 650
660 - 670 - 680 - 690 - 700 - 710 - 720 - 730 - 740 - 750 - 760 - 770 - 780
790 - 800 - 810 - 820 - 830 - 840 - 850 - 860 - 870 - 880 - 890 - 900 - 910
920 - 930 - 940 - 950 - 960 - 970 - 980 - 990 - 1000

Notes: ______________________________

____

____

____

Reg Reps: 10 - 20 - 30 - 40 - 50 - 60 - 70 - 80 - 90 - 100 - 110 - 120 - 130
140 - 150 - 160 - 170 - 180 - 190 - 200 - 210 - 220 - 230 - 240 - 250 - 260
270 - 280 - 290 - 300 - 310 - 320 - 330 - 340 - 350 - 360 - 370 - 380 - 390
400 - 410 - 420 - 430 - 440 - 450 - 460 - 470 - 480 - 490 - 500 - 510 - 520
530 - 540 - 550 - 560 - 570 - 580 - 590 - 600 - 610 - 620 - 630 - 640 - 650
660 - 670 - 680 - 690 - 700 - 710 - 720 - 730 - 740 - 750 - 760 - 770 - 780
790 - 800 - 810 - 820 - 830 - 840 - 850 - 860 - 870 - 880 - 890 - 900 - 910
920 - 930 - 940 - 950 - 960 - 970 - 980 - 990 - 1000

Notes: ______________________________

____

____

____

Reg Reps: 10 - 20 - 30 - 40 - 50 - 60 - 70 - 80 - 90 - 100 - 110 - 120 - 130
140 - 150 - 160 - 170 - 180 - 190 - 200 - 210 - 220 - 230 - 240 - 250 - 260
270 - 280 - 290 - 300 - 310 - 320 - 330 - 340 - 350 - 360 - 370 - 380 - 390
400 - 410 - 420 - 430 - 440 - 450 - 460 - 470 - 480 - 490 - 500 - 510 - 520
530 - 540 - 550 - 560 - 570 - 580 - 590 - 600 - 610 - 620 - 630 - 640 - 650
660 - 670 - 680 - 690 - 700 - 710 - 720 - 730 - 740 - 750 - 760 - 770 - 780
790 - 800 - 810 - 820 - 830 - 840 - 850 - 860 - 870 - 880 - 890 - 900 - 910
920 - 930 - 940 - 950 - 960 - 970 - 980 - 990 - 1000

Notes: ______________________________

____

____

______________________________ **End Date:** ____________

Defense Against:____________________________ **Start Date:** __________

Notes:____

Slow Reps: 10 - 20 - 30 - 40 - 50 - 60 - 70 - 80 - 90 - 100 - 110 - 120 - 130
140 - 150 - 160 - 170 - 180 - 190 - 200 - 210 - 220 - 230 - 240 - 250 - 260
270 - 280 - 290 - 300 - 310 - 320 - 330 - 340 - 350 - 360 - 370 - 380 - 390
400 - 410 - 420 - 430 - 440 - 450 - 460 - 470 - 480 - 490 - 500 - 510 - 520
530 - 540 - 550 - 560 - 570 - 580 - 590 - 600 - 610 - 620 - 630 - 640 - 650
660 - 670 - 680 - 690 - 700 - 710 - 720 - 730 - 740 - 750 - 760 - 770 - 780
790 - 800 - 810 - 820 - 830 - 840 - 850 - 860 - 870 - 880 - 890 - 900 - 910
920 - 930 - 940 - 950 - 960 - 970 - 980 - 990 - 1000

Reg Reps: 10 - 20 - 30 - 40 - 50 - 60 - 70 - 80 - 90 - 100 - 110 - 120 - 130
140 - 150 - 160 - 170 - 180 - 190 - 200 - 210 - 220 - 230 - 240 - 250 - 260
270 - 280 - 290 - 300 - 310 - 320 - 330 - 340 - 350 - 360 - 370 - 380 - 390
400 - 410 - 420 - 430 - 440 - 450 - 460 - 470 - 480 - 490 - 500 - 510 - 520
530 - 540 - 550 - 560 - 570 - 580 - 590 - 600 - 610 - 620 - 630 - 640 - 650
660 - 670 - 680 - 690 - 700 - 710 - 720 - 730 - 740 - 750 - 760 - 770 - 780
790 - 800 - 810 - 820 - 830 - 840 - 850 - 860 - 870 - 880 - 890 - 900 - 910
920 - 930 - 940 - 950 - 960 - 970 - 980 - 990 - 1000

Notes: ____

Reg Reps: 10 - 20 - 30 - 40 - 50 - 60 - 70 - 80 - 90 - 100 - 110 - 120 - 130
140 - 150 - 160 - 170 - 180 - 190 - 200 - 210 - 220 - 230 - 240 - 250 - 260
270 - 280 - 290 - 300 - 310 - 320 - 330 - 340 - 350 - 360 - 370 - 380 - 390
400 - 410 - 420 - 430 - 440 - 450 - 460 - 470 - 480 - 490 - 500 - 510 - 520
530 - 540 - 550 - 560 - 570 - 580 - 590 - 600 - 610 - 620 - 630 - 640 - 650
660 - 670 - 680 - 690 - 700 - 710 - 720 - 730 - 740 - 750 - 760 - 770 - 780
790 - 800 - 810 - 820 - 830 - 840 - 850 - 860 - 870 - 880 - 890 - 900 - 910
920 - 930 - 940 - 950 - 960 - 970 - 980 - 990 - 1000

Notes: ____

Reg Reps: 10 - 20 - 30 - 40 - 50 - 60 - 70 - 80 - 90 - 100 - 110 - 120 - 130
140 - 150 - 160 - 170 - 180 - 190 - 200 - 210 - 220 - 230 - 240 - 250 - 260
270 - 280 - 290 - 300 - 310 - 320 - 330 - 340 - 350 - 360 - 370 - 380 - 390
400 - 410 - 420 - 430 - 440 - 450 - 460 - 470 - 480 - 490 - 500 - 510 - 520
530 - 540 - 550 - 560 - 570 - 580 - 590 - 600 - 610 - 620 - 630 - 640 - 650
660 - 670 - 680 - 690 - 700 - 710 - 720 - 730 - 740 - 750 - 760 - 770 - 780
790 - 800 - 810 - 820 - 830 - 840 - 850 - 860 - 870 - 880 - 890 - 900 - 910
920 - 930 - 940 - 950 - 960 - 970 - 980 - 990 - 1000

Notes: ______________________________

____

____

____

Reg Reps: 10 - 20 - 30 - 40 - 50 - 60 - 70 - 80 - 90 - 100 - 110 - 120 - 130
140 - 150 - 160 - 170 - 180 - 190 - 200 - 210 - 220 - 230 - 240 - 250 - 260
270 - 280 - 290 - 300 - 310 - 320 - 330 - 340 - 350 - 360 - 370 - 380 - 390
400 - 410 - 420 - 430 - 440 - 450 - 460 - 470 - 480 - 490 - 500 - 510 - 520
530 - 540 - 550 - 560 - 570 - 580 - 590 - 600 - 610 - 620 - 630 - 640 - 650
660 - 670 - 680 - 690 - 700 - 710 - 720 - 730 - 740 - 750 - 760 - 770 - 780
790 - 800 - 810 - 820 - 830 - 840 - 850 - 860 - 870 - 880 - 890 - 900 - 910
920 - 930 - 940 - 950 - 960 - 970 - 980 - 990 - 1000

Notes: ______________________________

____

____

____

Reg Reps: 10 - 20 - 30 - 40 - 50 - 60 - 70 - 80 - 90 - 100 - 110 - 120 - 130
140 - 150 - 160 - 170 - 180 - 190 - 200 - 210 - 220 - 230 - 240 - 250 - 260
270 - 280 - 290 - 300 - 310 - 320 - 330 - 340 - 350 - 360 - 370 - 380 - 390
400 - 410 - 420 - 430 - 440 - 450 - 460 - 470 - 480 - 490 - 500 - 510 - 520
530 - 540 - 550 - 560 - 570 - 580 - 590 - 600 - 610 - 620 - 630 - 640 - 650
660 - 670 - 680 - 690 - 700 - 710 - 720 - 730 - 740 - 750 - 760 - 770 - 780
790 - 800 - 810 - 820 - 830 - 840 - 850 - 860 - 870 - 880 - 890 - 900 - 910
920 - 930 - 940 - 950 - 960 - 970 - 980 - 990 - 1000

Notes: ______________________________

____

____

______________________________ **End Date:** ____________

Defense Against:__________________________ **Start Date:** __________

Notes:____

____

____

____

Slow Reps: 10 - 20 - 30 - 40 - 50 - 60 - 70 - 80 - 90 - 100 - 110 - 120 - 130
140 - 150 - 160 - 170 - 180 - 190 - 200 - 210 - 220 - 230 - 240 - 250 - 260
270 - 280 - 290 - 300 - 310 - 320 - 330 - 340 - 350 - 360 - 370 - 380 - 390
400 - 410 - 420 - 430 - 440 - 450 - 460 - 470 - 480 - 490 - 500 - 510 - 520
530 - 540 - 550 - 560 - 570 - 580 - 590 - 600 - 610 - 620 - 630 - 640 - 650
660 - 670 - 680 - 690 - 700 - 710 - 720 - 730 - 740 - 750 - 760 - 770 - 780
790 - 800 - 810 - 820 - 830 - 840 - 850 - 860 - 870 - 880 - 890 - 900 - 910
920 - 930 - 940 - 950 - 960 - 970 - 980 - 990 - 1000

Reg Reps: 10 - 20 - 30 - 40 - 50 - 60 - 70 - 80 - 90 - 100 - 110 - 120 - 130
140 - 150 - 160 - 170 - 180 - 190 - 200 - 210 - 220 - 230 - 240 - 250 - 260
270 - 280 - 290 - 300 - 310 - 320 - 330 - 340 - 350 - 360 - 370 - 380 - 390
400 - 410 - 420 - 430 - 440 - 450 - 460 - 470 - 480 - 490 - 500 - 510 - 520
530 - 540 - 550 - 560 - 570 - 580 - 590 - 600 - 610 - 620 - 630 - 640 - 650
660 - 670 - 680 - 690 - 700 - 710 - 720 - 730 - 740 - 750 - 760 - 770 - 780
790 - 800 - 810 - 820 - 830 - 840 - 850 - 860 - 870 - 880 - 890 - 900 - 910
920 - 930 - 940 - 950 - 960 - 970 - 980 - 990 - 1000

Notes: ____

____

____

____

Reg Reps: 10 - 20 - 30 - 40 - 50 - 60 - 70 - 80 - 90 - 100 - 110 - 120 - 130
140 - 150 - 160 - 170 - 180 - 190 - 200 - 210 - 220 - 230 - 240 - 250 - 260
270 - 280 - 290 - 300 - 310 - 320 - 330 - 340 - 350 - 360 - 370 - 380 - 390
400 - 410 - 420 - 430 - 440 - 450 - 460 - 470 - 480 - 490 - 500 - 510 - 520
530 - 540 - 550 - 560 - 570 - 580 - 590 - 600 - 610 - 620 - 630 - 640 - 650
660 - 670 - 680 - 690 - 700 - 710 - 720 - 730 - 740 - 750 - 760 - 770 - 780
790 - 800 - 810 - 820 - 830 - 840 - 850 - 860 - 870 - 880 - 890 - 900 - 910
920 - 930 - 940 - 950 - 960 - 970 - 980 - 990 - 1000

Notes: ____

____

____

____

Reg Reps: 10 - 20 - 30 - 40 - 50 - 60 - 70 - 80 - 90 - 100 - 110 - 120 - 130
140 - 150 - 160 - 170 - 180 - 190 - 200 - 210 - 220 - 230 - 240 - 250 - 260
270 - 280 - 290 - 300 - 310 - 320 - 330 - 340 - 350 - 360 - 370 - 380 - 390
400 - 410 - 420 - 430 - 440 - 450 - 460 - 470 - 480 - 490 - 500 - 510 - 520
530 - 540 - 550 - 560 - 570 - 580 - 590 - 600 - 610 - 620 - 630 - 640 - 650
660 - 670 - 680 - 690 - 700 - 710 - 720 - 730 - 740 - 750 - 760 - 770 - 780
790 - 800 - 810 - 820 - 830 - 840 - 850 - 860 - 870 - 880 - 890 - 900 - 910
920 - 930 - 940 - 950 - 960 - 970 - 980 - 990 - 1000

Notes: ______________________________

____

____

____

Reg Reps: 10 - 20 - 30 - 40 - 50 - 60 - 70 - 80 - 90 - 100 - 110 - 120 - 130
140 - 150 - 160 - 170 - 180 - 190 - 200 - 210 - 220 - 230 - 240 - 250 - 260
270 - 280 - 290 - 300 - 310 - 320 - 330 - 340 - 350 - 360 - 370 - 380 - 390
400 - 410 - 420 - 430 - 440 - 450 - 460 - 470 - 480 - 490 - 500 - 510 - 520
530 - 540 - 550 - 560 - 570 - 580 - 590 - 600 - 610 - 620 - 630 - 640 - 650
660 - 670 - 680 - 690 - 700 - 710 - 720 - 730 - 740 - 750 - 760 - 770 - 780
790 - 800 - 810 - 820 - 830 - 840 - 850 - 860 - 870 - 880 - 890 - 900 - 910
920 - 930 - 940 - 950 - 960 - 970 - 980 - 990 - 1000

Notes: ______________________________

____

____

____

Reg Reps: 10 - 20 - 30 - 40 - 50 - 60 - 70 - 80 - 90 - 100 - 110 - 120 - 130
140 - 150 - 160 - 170 - 180 - 190 - 200 - 210 - 220 - 230 - 240 - 250 - 260
270 - 280 - 290 - 300 - 310 - 320 - 330 - 340 - 350 - 360 - 370 - 380 - 390
400 - 410 - 420 - 430 - 440 - 450 - 460 - 470 - 480 - 490 - 500 - 510 - 520
530 - 540 - 550 - 560 - 570 - 580 - 590 - 600 - 610 - 620 - 630 - 640 - 650
660 - 670 - 680 - 690 - 700 - 710 - 720 - 730 - 740 - 750 - 760 - 770 - 780
790 - 800 - 810 - 820 - 830 - 840 - 850 - 860 - 870 - 880 - 890 - 900 - 910
920 - 930 - 940 - 950 - 960 - 970 - 980 - 990 - 1000

Notes: ______________________________

____

____

______________________________ **End Date:** ____________

Defense Against:____________________________ **Start Date:** __________

Notes:____

____

____

____

Slow Reps: 10 - 20 - 30 - 40 - 50 - 60 - 70 - 80 - 90 - 100 - 110 - 120 - 130
140 - 150 - 160 - 170 - 180 - 190 - 200 - 210 - 220 - 230 - 240 - 250 - 260
270 - 280 - 290 - 300 - 310 - 320 - 330 - 340 - 350 - 360 - 370 - 380 - 390
400 - 410 - 420 - 430 - 440 - 450 - 460 - 470 - 480 - 490 - 500 - 510 - 520
530 - 540 - 550 - 560 - 570 - 580 - 590 - 600 - 610 - 620 - 630 - 640 - 650
660 - 670 - 680 - 690 - 700 - 710 - 720 - 730 - 740 - 750 - 760 - 770 - 780
790 - 800 - 810 - 820 - 830 - 840 - 850 - 860 - 870 - 880 - 890 - 900 - 910
920 - 930 - 940 - 950 - 960 - 970 - 980 - 990 - 1000

Reg Reps: 10 - 20 - 30 - 40 - 50 - 60 - 70 - 80 - 90 - 100 - 110 - 120 - 130
140 - 150 - 160 - 170 - 180 - 190 - 200 - 210 - 220 - 230 - 240 - 250 - 260
270 - 280 - 290 - 300 - 310 - 320 - 330 - 340 - 350 - 360 - 370 - 380 - 390
400 - 410 - 420 - 430 - 440 - 450 - 460 - 470 - 480 - 490 - 500 - 510 - 520
530 - 540 - 550 - 560 - 570 - 580 - 590 - 600 - 610 - 620 - 630 - 640 - 650
660 - 670 - 680 - 690 - 700 - 710 - 720 - 730 - 740 - 750 - 760 - 770 - 780
790 - 800 - 810 - 820 - 830 - 840 - 850 - 860 - 870 - 880 - 890 - 900 - 910
920 - 930 - 940 - 950 - 960 - 970 - 980 - 990 - 1000

Notes: ____

____

____

____

Reg Reps: 10 - 20 - 30 - 40 - 50 - 60 - 70 - 80 - 90 - 100 - 110 - 120 - 130
140 - 150 - 160 - 170 - 180 - 190 - 200 - 210 - 220 - 230 - 240 - 250 - 260
270 - 280 - 290 - 300 - 310 - 320 - 330 - 340 - 350 - 360 - 370 - 380 - 390
400 - 410 - 420 - 430 - 440 - 450 - 460 - 470 - 480 - 490 - 500 - 510 - 520
530 - 540 - 550 - 560 - 570 - 580 - 590 - 600 - 610 - 620 - 630 - 640 - 650
660 - 670 - 680 - 690 - 700 - 710 - 720 - 730 - 740 - 750 - 760 - 770 - 780
790 - 800 - 810 - 820 - 830 - 840 - 850 - 860 - 870 - 880 - 890 - 900 - 910
920 - 930 - 940 - 950 - 960 - 970 - 980 - 990 - 1000

Notes: ____

____

____

____

Reg Reps: 10 - 20 - 30 - 40 - 50 - 60 - 70 - 80 - 90 - 100 - 110 - 120 - 130
140 - 150 - 160 - 170 - 180 - 190 - 200 - 210 - 220 - 230 - 240 - 250 - 260
270 - 280 - 290 - 300 - 310 - 320 - 330 - 340 - 350 - 360 - 370 - 380 - 390
400 - 410 - 420 - 430 - 440 - 450 - 460 - 470 - 480 - 490 - 500 - 510 - 520
530 - 540 - 550 - 560 - 570 - 580 - 590 - 600 - 610 - 620 - 630 - 640 - 650
660 - 670 - 680 - 690 - 700 - 710 - 720 - 730 - 740 - 750 - 760 - 770 - 780
790 - 800 - 810 - 820 - 830 - 840 - 850 - 860 - 870 - 880 - 890 - 900 - 910
920 - 930 - 940 - 950 - 960 - 970 - 980 - 990 - 1000

Notes: ______________________________

Reg Reps: 10 - 20 - 30 - 40 - 50 - 60 - 70 - 80 - 90 - 100 - 110 - 120 - 130
140 - 150 - 160 - 170 - 180 - 190 - 200 - 210 - 220 - 230 - 240 - 250 - 260
270 - 280 - 290 - 300 - 310 - 320 - 330 - 340 - 350 - 360 - 370 - 380 - 390
400 - 410 - 420 - 430 - 440 - 450 - 460 - 470 - 480 - 490 - 500 - 510 - 520
530 - 540 - 550 - 560 - 570 - 580 - 590 - 600 - 610 - 620 - 630 - 640 - 650
660 - 670 - 680 - 690 - 700 - 710 - 720 - 730 - 740 - 750 - 760 - 770 - 780
790 - 800 - 810 - 820 - 830 - 840 - 850 - 860 - 870 - 880 - 890 - 900 - 910
920 - 930 - 940 - 950 - 960 - 970 - 980 - 990 - 1000

Notes: ______________________________

Reg Reps: 10 - 20 - 30 - 40 - 50 - 60 - 70 - 80 - 90 - 100 - 110 - 120 - 130
140 - 150 - 160 - 170 - 180 - 190 - 200 - 210 - 220 - 230 - 240 - 250 - 260
270 - 280 - 290 - 300 - 310 - 320 - 330 - 340 - 350 - 360 - 370 - 380 - 390
400 - 410 - 420 - 430 - 440 - 450 - 460 - 470 - 480 - 490 - 500 - 510 - 520
530 - 540 - 550 - 560 - 570 - 580 - 590 - 600 - 610 - 620 - 630 - 640 - 650
660 - 670 - 680 - 690 - 700 - 710 - 720 - 730 - 740 - 750 - 760 - 770 - 780
790 - 800 - 810 - 820 - 830 - 840 - 850 - 860 - 870 - 880 - 890 - 900 - 910
920 - 930 - 940 - 950 - 960 - 970 - 980 - 990 - 1000

Notes: ______________________________

End Date: ______________

Defense Against:____________________________ **Start Date:** __________

Notes:____

____

____

____

Slow Reps: 10 - 20 - 30 - 40 - 50 - 60 - 70 - 80 - 90 - 100 - 110 - 120 - 130
140 - 150 - 160 - 170 - 180 - 190 - 200 - 210 - 220 - 230 - 240 - 250 - 260
270 - 280 - 290 - 300 - 310 - 320 - 330 - 340 - 350 - 360 - 370 - 380 - 390
400 - 410 - 420 - 430 - 440 - 450 - 460 - 470 - 480 - 490 - 500 - 510 - 520
530 - 540 - 550 - 560 - 570 - 580 - 590 - 600 - 610 - 620 - 630 - 640 - 650
660 - 670 - 680 - 690 - 700 - 710 - 720 - 730 - 740 - 750 - 760 - 770 - 780
790 - 800 - 810 - 820 - 830 - 840 - 850 - 860 - 870 - 880 - 890 - 900 - 910
920 - 930 - 940 - 950 - 960 - 970 - 980 - 990 - 1000

Reg Reps: 10 - 20 - 30 - 40 - 50 - 60 - 70 - 80 - 90 - 100 - 110 - 120 - 130
140 - 150 - 160 - 170 - 180 - 190 - 200 - 210 - 220 - 230 - 240 - 250 - 260
270 - 280 - 290 - 300 - 310 - 320 - 330 - 340 - 350 - 360 - 370 - 380 - 390
400 - 410 - 420 - 430 - 440 - 450 - 460 - 470 - 480 - 490 - 500 - 510 - 520
530 - 540 - 550 - 560 - 570 - 580 - 590 - 600 - 610 - 620 - 630 - 640 - 650
660 - 670 - 680 - 690 - 700 - 710 - 720 - 730 - 740 - 750 - 760 - 770 - 780
790 - 800 - 810 - 820 - 830 - 840 - 850 - 860 - 870 - 880 - 890 - 900 - 910
920 - 930 - 940 - 950 - 960 - 970 - 980 - 990 - 1000

Notes: ____

____

____

____

Reg Reps: 10 - 20 - 30 - 40 - 50 - 60 - 70 - 80 - 90 - 100 - 110 - 120 - 130
140 - 150 - 160 - 170 - 180 - 190 - 200 - 210 - 220 - 230 - 240 - 250 - 260
270 - 280 - 290 - 300 - 310 - 320 - 330 - 340 - 350 - 360 - 370 - 380 - 390
400 - 410 - 420 - 430 - 440 - 450 - 460 - 470 - 480 - 490 - 500 - 510 - 520
530 - 540 - 550 - 560 - 570 - 580 - 590 - 600 - 610 - 620 - 630 - 640 - 650
660 - 670 - 680 - 690 - 700 - 710 - 720 - 730 - 740 - 750 - 760 - 770 - 780
790 - 800 - 810 - 820 - 830 - 840 - 850 - 860 - 870 - 880 - 890 - 900 - 910
920 - 930 - 940 - 950 - 960 - 970 - 980 - 990 - 1000

Notes: ____

____

____

____

Reg Reps: 10 - 20 - 30 - 40 - 50 - 60 - 70 - 80 - 90 - 100 - 110 - 120 - 130
140 - 150 - 160 - 170 - 180 - 190 - 200 - 210 - 220 - 230 - 240 - 250 - 260
270 - 280 - 290 - 300 - 310 - 320 - 330 - 340 - 350 - 360 - 370 - 380 - 390
400 - 410 - 420 - 430 - 440 - 450 - 460 - 470 - 480 - 490 - 500 - 510 - 520
530 - 540 - 550 - 560 - 570 - 580 - 590 - 600 - 610 - 620 - 630 - 640 - 650
660 - 670 - 680 - 690 - 700 - 710 - 720 - 730 - 740 - 750 - 760 - 770 - 780
790 - 800 - 810 - 820 - 830 - 840 - 850 - 860 - 870 - 880 - 890 - 900 - 910
920 - 930 - 940 - 950 - 960 - 970 - 980 - 990 - 1000

Notes: ____________________

Reg Reps: 10 - 20 - 30 - 40 - 50 - 60 - 70 - 80 - 90 - 100 - 110 - 120 - 130
140 - 150 - 160 - 170 - 180 - 190 - 200 - 210 - 220 - 230 - 240 - 250 - 260
270 - 280 - 290 - 300 - 310 - 320 - 330 - 340 - 350 - 360 - 370 - 380 - 390
400 - 410 - 420 - 430 - 440 - 450 - 460 - 470 - 480 - 490 - 500 - 510 - 520
530 - 540 - 550 - 560 - 570 - 580 - 590 - 600 - 610 - 620 - 630 - 640 - 650
660 - 670 - 680 - 690 - 700 - 710 - 720 - 730 - 740 - 750 - 760 - 770 - 780
790 - 800 - 810 - 820 - 830 - 840 - 850 - 860 - 870 - 880 - 890 - 900 - 910
920 - 930 - 940 - 950 - 960 - 970 - 980 - 990 - 1000

Notes: ____________________

Reg Reps: 10 - 20 - 30 - 40 - 50 - 60 - 70 - 80 - 90 - 100 - 110 - 120 - 130
140 - 150 - 160 - 170 - 180 - 190 - 200 - 210 - 220 - 230 - 240 - 250 - 260
270 - 280 - 290 - 300 - 310 - 320 - 330 - 340 - 350 - 360 - 370 - 380 - 390
400 - 410 - 420 - 430 - 440 - 450 - 460 - 470 - 480 - 490 - 500 - 510 - 520
530 - 540 - 550 - 560 - 570 - 580 - 590 - 600 - 610 - 620 - 630 - 640 - 650
660 - 670 - 680 - 690 - 700 - 710 - 720 - 730 - 740 - 750 - 760 - 770 - 780
790 - 800 - 810 - 820 - 830 - 840 - 850 - 860 - 870 - 880 - 890 - 900 - 910
920 - 930 - 940 - 950 - 960 - 970 - 980 - 990 - 1000

Notes: ____________________

End Date: ____________

Defense Against:________________________ **Start Date:** _________

Notes:____

Slow Reps: 10 - 20 - 30 - 40 - 50 - 60 - 70 - 80 - 90 - 100 - 110 - 120 - 130
140 - 150 - 160 - 170 - 180 - 190 - 200 - 210 - 220 - 230 - 240 - 250 - 260
270 - 280 - 290 - 300 - 310 - 320 - 330 - 340 - 350 - 360 - 370 - 380 - 390
400 - 410 - 420 - 430 - 440 - 450 - 460 - 470 - 480 - 490 - 500 - 510 - 520
530 - 540 - 550 - 560 - 570 - 580 - 590 - 600 - 610 - 620 - 630 - 640 - 650
660 - 670 - 680 - 690 - 700 - 710 - 720 - 730 - 740 - 750 - 760 - 770 - 780
790 - 800 - 810 - 820 - 830 - 840 - 850 - 860 - 870 - 880 - 890 - 900 - 910
920 - 930 - 940 - 950 - 960 - 970 - 980 - 990 - 1000

Reg Reps: 10 - 20 - 30 - 40 - 50 - 60 - 70 - 80 - 90 - 100 - 110 - 120 - 130
140 - 150 - 160 - 170 - 180 - 190 - 200 - 210 - 220 - 230 - 240 - 250 - 260
270 - 280 - 290 - 300 - 310 - 320 - 330 - 340 - 350 - 360 - 370 - 380 - 390
400 - 410 - 420 - 430 - 440 - 450 - 460 - 470 - 480 - 490 - 500 - 510 - 520
530 - 540 - 550 - 560 - 570 - 580 - 590 - 600 - 610 - 620 - 630 - 640 - 650
660 - 670 - 680 - 690 - 700 - 710 - 720 - 730 - 740 - 750 - 760 - 770 - 780
790 - 800 - 810 - 820 - 830 - 840 - 850 - 860 - 870 - 880 - 890 - 900 - 910
920 - 930 - 940 - 950 - 960 - 970 - 980 - 990 - 1000

Notes: ____

Reg Reps: 10 - 20 - 30 - 40 - 50 - 60 - 70 - 80 - 90 - 100 - 110 - 120 - 130
140 - 150 - 160 - 170 - 180 - 190 - 200 - 210 - 220 - 230 - 240 - 250 - 260
270 - 280 - 290 - 300 - 310 - 320 - 330 - 340 - 350 - 360 - 370 - 380 - 390
400 - 410 - 420 - 430 - 440 - 450 - 460 - 470 - 480 - 490 - 500 - 510 - 520
530 - 540 - 550 - 560 - 570 - 580 - 590 - 600 - 610 - 620 - 630 - 640 - 650
660 - 670 - 680 - 690 - 700 - 710 - 720 - 730 - 740 - 750 - 760 - 770 - 780
790 - 800 - 810 - 820 - 830 - 840 - 850 - 860 - 870 - 880 - 890 - 900 - 910
920 - 930 - 940 - 950 - 960 - 970 - 980 - 990 - 1000

Notes: ____

Reg Reps: 10 - 20 - 30 - 40 - 50 - 60 - 70 - 80 - 90 - 100 - 110 - 120 - 130 - 140 - 150 - 160 - 170 - 180 - 190 - 200 - 210 - 220 - 230 - 240 - 250 - 260 - 270 - 280 - 290 - 300 - 310 - 320 - 330 - 340 - 350 - 360 - 370 - 380 - 390 - 400 - 410 - 420 - 430 - 440 - 450 - 460 - 470 - 480 - 490 - 500 - 510 - 520 - 530 - 540 - 550 - 560 - 570 - 580 - 590 - 600 - 610 - 620 - 630 - 640 - 650 - 660 - 670 - 680 - 690 - 700 - 710 - 720 - 730 - 740 - 750 - 760 - 770 - 780 - 790 - 800 - 810 - 820 - 830 - 840 - 850 - 860 - 870 - 880 - 890 - 900 - 910 - 920 - 930 - 940 - 950 - 960 - 970 - 980 - 990 - 1000

Notes: ______________________________

____

____

____

Reg Reps: 10 - 20 - 30 - 40 - 50 - 60 - 70 - 80 - 90 - 100 - 110 - 120 - 130 - 140 - 150 - 160 - 170 - 180 - 190 - 200 - 210 - 220 - 230 - 240 - 250 - 260 - 270 - 280 - 290 - 300 - 310 - 320 - 330 - 340 - 350 - 360 - 370 - 380 - 390 - 400 - 410 - 420 - 430 - 440 - 450 - 460 - 470 - 480 - 490 - 500 - 510 - 520 - 530 - 540 - 550 - 560 - 570 - 580 - 590 - 600 - 610 - 620 - 630 - 640 - 650 - 660 - 670 - 680 - 690 - 700 - 710 - 720 - 730 - 740 - 750 - 760 - 770 - 780 - 790 - 800 - 810 - 820 - 830 - 840 - 850 - 860 - 870 - 880 - 890 - 900 - 910 - 920 - 930 - 940 - 950 - 960 - 970 - 980 - 990 - 1000

Notes: ______________________________

____

____

____

Reg Reps: 10 - 20 - 30 - 40 - 50 - 60 - 70 - 80 - 90 - 100 - 110 - 120 - 130 - 140 - 150 - 160 - 170 - 180 - 190 - 200 - 210 - 220 - 230 - 240 - 250 - 260 - 270 - 280 - 290 - 300 - 310 - 320 - 330 - 340 - 350 - 360 - 370 - 380 - 390 - 400 - 410 - 420 - 430 - 440 - 450 - 460 - 470 - 480 - 490 - 500 - 510 - 520 - 530 - 540 - 550 - 560 - 570 - 580 - 590 - 600 - 610 - 620 - 630 - 640 - 650 - 660 - 670 - 680 - 690 - 700 - 710 - 720 - 730 - 740 - 750 - 760 - 770 - 780 - 790 - 800 - 810 - 820 - 830 - 840 - 850 - 860 - 870 - 880 - 890 - 900 - 910 - 920 - 930 - 940 - 950 - 960 - 970 - 980 - 990 - 1000

Notes: ______________________________

____

____

______________________________ **End Date:** ______________

Defense Against:____________________________ **Start Date:** __________

Notes:____

Slow Reps: 10 - 20 - 30 - 40 - 50 - 60 - 70 - 80 - 90 - 100 - 110 - 120 - 130
140 - 150 - 160 - 170 - 180 - 190 - 200 - 210 - 220 - 230 - 240 - 250 - 260
270 - 280 - 290 - 300 - 310 - 320 - 330 - 340 - 350 - 360 - 370 - 380 - 390
400 - 410 - 420 - 430 - 440 - 450 - 460 - 470 - 480 - 490 - 500 - 510 - 520
530 - 540 - 550 - 560 - 570 - 580 - 590 - 600 - 610 - 620 - 630 - 640 - 650
660 - 670 - 680 - 690 - 700 - 710 - 720 - 730 - 740 - 750 - 760 - 770 - 780
790 - 800 - 810 - 820 - 830 - 840 - 850 - 860 - 870 - 880 - 890 - 900 - 910
920 - 930 - 940 - 950 - 960 - 970 - 980 - 990 - 1000

Reg Reps: 10 - 20 - 30 - 40 - 50 - 60 - 70 - 80 - 90 - 100 - 110 - 120 - 130
140 - 150 - 160 - 170 - 180 - 190 - 200 - 210 - 220 - 230 - 240 - 250 - 260
270 - 280 - 290 - 300 - 310 - 320 - 330 - 340 - 350 - 360 - 370 - 380 - 390
400 - 410 - 420 - 430 - 440 - 450 - 460 - 470 - 480 - 490 - 500 - 510 - 520
530 - 540 - 550 - 560 - 570 - 580 - 590 - 600 - 610 - 620 - 630 - 640 - 650
660 - 670 - 680 - 690 - 700 - 710 - 720 - 730 - 740 - 750 - 760 - 770 - 780
790 - 800 - 810 - 820 - 830 - 840 - 850 - 860 - 870 - 880 - 890 - 900 - 910
920 - 930 - 940 - 950 - 960 - 970 - 980 - 990 - 1000

Notes: ____

Reg Reps: 10 - 20 - 30 - 40 - 50 - 60 - 70 - 80 - 90 - 100 - 110 - 120 - 130
140 - 150 - 160 - 170 - 180 - 190 - 200 - 210 - 220 - 230 - 240 - 250 - 260
270 - 280 - 290 - 300 - 310 - 320 - 330 - 340 - 350 - 360 - 370 - 380 - 390
400 - 410 - 420 - 430 - 440 - 450 - 460 - 470 - 480 - 490 - 500 - 510 - 520
530 - 540 - 550 - 560 - 570 - 580 - 590 - 600 - 610 - 620 - 630 - 640 - 650
660 - 670 - 680 - 690 - 700 - 710 - 720 - 730 - 740 - 750 - 760 - 770 - 780
790 - 800 - 810 - 820 - 830 - 840 - 850 - 860 - 870 - 880 - 890 - 900 - 910
920 - 930 - 940 - 950 - 960 - 970 - 980 - 990 - 1000

Notes: ____

Reg Reps: 10 - 20 - 30 - 40 - 50 - 60 - 70 - 80 - 90 - 100 - 110 - 120 - 130 140 - 150 - 160 - 170 - 180 - 190 - 200 - 210 - 220 - 230 - 240 - 250 - 260 270 - 280 - 290 - 300 - 310 - 320 - 330 - 340 - 350 - 360 - 370 - 380 - 390 400 - 410 - 420 - 430 - 440 - 450 - 460 - 470 - 480 - 490 - 500 - 510 - 520 530 - 540 - 550 - 560 - 570 - 580 - 590 - 600 - 610 - 620 - 630 - 640 - 650 660 - 670 - 680 - 690 - 700 - 710 - 720 - 730 - 740 - 750 - 760 - 770 - 780 790 - 800 - 810 - 820 - 830 - 840 - 850 - 860 - 870 - 880 - 890 - 900 - 910 920 - 930 - 940 - 950 - 960 - 970 - 980 - 990 - 1000

Notes: ______________________________

____

____

____

Reg Reps: 10 - 20 - 30 - 40 - 50 - 60 - 70 - 80 - 90 - 100 - 110 - 120 - 130 140 - 150 - 160 - 170 - 180 - 190 - 200 - 210 - 220 - 230 - 240 - 250 - 260 270 - 280 - 290 - 300 - 310 - 320 - 330 - 340 - 350 - 360 - 370 - 380 - 390 400 - 410 - 420 - 430 - 440 - 450 - 460 - 470 - 480 - 490 - 500 - 510 - 520 530 - 540 - 550 - 560 - 570 - 580 - 590 - 600 - 610 - 620 - 630 - 640 - 650 660 - 670 - 680 - 690 - 700 - 710 - 720 - 730 - 740 - 750 - 760 - 770 - 780 790 - 800 - 810 - 820 - 830 - 840 - 850 - 860 - 870 - 880 - 890 - 900 - 910 920 - 930 - 940 - 950 - 960 - 970 - 980 - 990 - 1000

Notes: ______________________________

____

____

____

Reg Reps: 10 - 20 - 30 - 40 - 50 - 60 - 70 - 80 - 90 - 100 - 110 - 120 - 130 140 - 150 - 160 - 170 - 180 - 190 - 200 - 210 - 220 - 230 - 240 - 250 - 260 270 - 280 - 290 - 300 - 310 - 320 - 330 - 340 - 350 - 360 - 370 - 380 - 390 400 - 410 - 420 - 430 - 440 - 450 - 460 - 470 - 480 - 490 - 500 - 510 - 520 530 - 540 - 550 - 560 - 570 - 580 - 590 - 600 - 610 - 620 - 630 - 640 - 650 660 - 670 - 680 - 690 - 700 - 710 - 720 - 730 - 740 - 750 - 760 - 770 - 780 790 - 800 - 810 - 820 - 830 - 840 - 850 - 860 - 870 - 880 - 890 - 900 - 910 920 - 930 - 940 - 950 - 960 - 970 - 980 - 990 - 1000

Notes: ______________________________

____

____

______________________________ **End Date:** ____________

Defense Against: ______________________ **Start Date:** __________

Notes: ____

____

____

____

Slow Reps: 10 - 20 - 30 - 40 - 50 - 60 - 70 - 80 - 90 - 100 - 110 - 120 - 130
140 - 150 - 160 - 170 - 180 - 190 - 200 - 210 - 220 - 230 - 240 - 250 - 260
270 - 280 - 290 - 300 - 310 - 320 - 330 - 340 - 350 - 360 - 370 - 380 - 390
400 - 410 - 420 - 430 - 440 - 450 - 460 - 470 - 480 - 490 - 500 - 510 - 520
530 - 540 - 550 - 560 - 570 - 580 - 590 - 600 - 610 - 620 - 630 - 640 - 650
660 - 670 - 680 - 690 - 700 - 710 - 720 - 730 - 740 - 750 - 760 - 770 - 780
790 - 800 - 810 - 820 - 830 - 840 - 850 - 860 - 870 - 880 - 890 - 900 - 910
920 - 930 - 940 - 950 - 960 - 970 - 980 - 990 - 1000

Reg Reps: 10 - 20 - 30 - 40 - 50 - 60 - 70 - 80 - 90 - 100 - 110 - 120 - 130
140 - 150 - 160 - 170 - 180 - 190 - 200 - 210 - 220 - 230 - 240 - 250 - 260
270 - 280 - 290 - 300 - 310 - 320 - 330 - 340 - 350 - 360 - 370 - 380 - 390
400 - 410 - 420 - 430 - 440 - 450 - 460 - 470 - 480 - 490 - 500 - 510 - 520
530 - 540 - 550 - 560 - 570 - 580 - 590 - 600 - 610 - 620 - 630 - 640 - 650
660 - 670 - 680 - 690 - 700 - 710 - 720 - 730 - 740 - 750 - 760 - 770 - 780
790 - 800 - 810 - 820 - 830 - 840 - 850 - 860 - 870 - 880 - 890 - 900 - 910
920 - 930 - 940 - 950 - 960 - 970 - 980 - 990 - 1000

Notes: ____

____

____

____

Reg Reps: 10 - 20 - 30 - 40 - 50 - 60 - 70 - 80 - 90 - 100 - 110 - 120 - 130
140 - 150 - 160 - 170 - 180 - 190 - 200 - 210 - 220 - 230 - 240 - 250 - 260
270 - 280 - 290 - 300 - 310 - 320 - 330 - 340 - 350 - 360 - 370 - 380 - 390
400 - 410 - 420 - 430 - 440 - 450 - 460 - 470 - 480 - 490 - 500 - 510 - 520
530 - 540 - 550 - 560 - 570 - 580 - 590 - 600 - 610 - 620 - 630 - 640 - 650
660 - 670 - 680 - 690 - 700 - 710 - 720 - 730 - 740 - 750 - 760 - 770 - 780
790 - 800 - 810 - 820 - 830 - 840 - 850 - 860 - 870 - 880 - 890 - 900 - 910
920 - 930 - 940 - 950 - 960 - 970 - 980 - 990 - 1000

Notes: ____

____

____

____

Reg Reps: 10 - 20 - 30 - 40 - 50 - 60 - 70 - 80 - 90 - 100 - 110 - 120 - 130
140 - 150 - 160 - 170 - 180 - 190 - 200 - 210 - 220 - 230 - 240 - 250 - 260
270 - 280 - 290 - 300 - 310 - 320 - 330 - 340 - 350 - 360 - 370 - 380 - 390
400 - 410 - 420 - 430 - 440 - 450 - 460 - 470 - 480 - 490 - 500 - 510 - 520
530 - 540 - 550 - 560 - 570 - 580 - 590 - 600 - 610 - 620 - 630 - 640 - 650
660 - 670 - 680 - 690 - 700 - 710 - 720 - 730 - 740 - 750 - 760 - 770 - 780
790 - 800 - 810 - 820 - 830 - 840 - 850 - 860 - 870 - 880 - 890 - 900 - 910
920 - 930 - 940 - 950 - 960 - 970 - 980 - 990 - 1000

Notes: ______________________________

____

____

____

Reg Reps: 10 - 20 - 30 - 40 - 50 - 60 - 70 - 80 - 90 - 100 - 110 - 120 - 130
140 - 150 - 160 - 170 - 180 - 190 - 200 - 210 - 220 - 230 - 240 - 250 - 260
270 - 280 - 290 - 300 - 310 - 320 - 330 - 340 - 350 - 360 - 370 - 380 - 390
400 - 410 - 420 - 430 - 440 - 450 - 460 - 470 - 480 - 490 - 500 - 510 - 520
530 - 540 - 550 - 560 - 570 - 580 - 590 - 600 - 610 - 620 - 630 - 640 - 650
660 - 670 - 680 - 690 - 700 - 710 - 720 - 730 - 740 - 750 - 760 - 770 - 780
790 - 800 - 810 - 820 - 830 - 840 - 850 - 860 - 870 - 880 - 890 - 900 - 910
920 - 930 - 940 - 950 - 960 - 970 - 980 - 990 - 1000

Notes: ______________________________

____

____

____

Reg Reps: 10 - 20 - 30 - 40 - 50 - 60 - 70 - 80 - 90 - 100 - 110 - 120 - 130
140 - 150 - 160 - 170 - 180 - 190 - 200 - 210 - 220 - 230 - 240 - 250 - 260
270 - 280 - 290 - 300 - 310 - 320 - 330 - 340 - 350 - 360 - 370 - 380 - 390
400 - 410 - 420 - 430 - 440 - 450 - 460 - 470 - 480 - 490 - 500 - 510 - 520
530 - 540 - 550 - 560 - 570 - 580 - 590 - 600 - 610 - 620 - 630 - 640 - 650
660 - 670 - 680 - 690 - 700 - 710 - 720 - 730 - 740 - 750 - 760 - 770 - 780
790 - 800 - 810 - 820 - 830 - 840 - 850 - 860 - 870 - 880 - 890 - 900 - 910
920 - 930 - 940 - 950 - 960 - 970 - 980 - 990 - 1000

Notes: ______________________________

____

____

______________________________ **End Date:** ____________

Defense Against:__________________________ **Start Date:** __________

Notes:____

____

____

____

Slow Reps: 10 - 20 - 30 - 40 - 50 - 60 - 70 - 80 - 90 - 100 - 110 - 120 - 130
140 - 150 - 160 - 170 - 180 - 190 - 200 - 210 - 220 - 230 - 240 - 250 - 260
270 - 280 - 290 - 300 - 310 - 320 - 330 - 340 - 350 - 360 - 370 - 380 - 390
400 - 410 - 420 - 430 - 440 - 450 - 460 - 470 - 480 - 490 - 500 - 510 - 520
530 - 540 - 550 - 560 - 570 - 580 - 590 - 600 - 610 - 620 - 630 - 640 - 650
660 - 670 - 680 - 690 - 700 - 710 - 720 - 730 - 740 - 750 - 760 - 770 - 780
790 - 800 - 810 - 820 - 830 - 840 - 850 - 860 - 870 - 880 - 890 - 900 - 910
920 - 930 - 940 - 950 - 960 - 970 - 980 - 990 - 1000

Reg Reps: 10 - 20 - 30 - 40 - 50 - 60 - 70 - 80 - 90 - 100 - 110 - 120 - 130
140 - 150 - 160 - 170 - 180 - 190 - 200 - 210 - 220 - 230 - 240 - 250 - 260
270 - 280 - 290 - 300 - 310 - 320 - 330 - 340 - 350 - 360 - 370 - 380 - 390
400 - 410 - 420 - 430 - 440 - 450 - 460 - 470 - 480 - 490 - 500 - 510 - 520
530 - 540 - 550 - 560 - 570 - 580 - 590 - 600 - 610 - 620 - 630 - 640 - 650
660 - 670 - 680 - 690 - 700 - 710 - 720 - 730 - 740 - 750 - 760 - 770 - 780
790 - 800 - 810 - 820 - 830 - 840 - 850 - 860 - 870 - 880 - 890 - 900 - 910
920 - 930 - 940 - 950 - 960 - 970 - 980 - 990 - 1000

Notes: ____

____

____

____

Reg Reps: 10 - 20 - 30 - 40 - 50 - 60 - 70 - 80 - 90 - 100 - 110 - 120 - 130
140 - 150 - 160 - 170 - 180 - 190 - 200 - 210 - 220 - 230 - 240 - 250 - 260
270 - 280 - 290 - 300 - 310 - 320 - 330 - 340 - 350 - 360 - 370 - 380 - 390
400 - 410 - 420 - 430 - 440 - 450 - 460 - 470 - 480 - 490 - 500 - 510 - 520
530 - 540 - 550 - 560 - 570 - 580 - 590 - 600 - 610 - 620 - 630 - 640 - 650
660 - 670 - 680 - 690 - 700 - 710 - 720 - 730 - 740 - 750 - 760 - 770 - 780
790 - 800 - 810 - 820 - 830 - 840 - 850 - 860 - 870 - 880 - 890 - 900 - 910
920 - 930 - 940 - 950 - 960 - 970 - 980 - 990 - 1000

Notes: ____

____

____

____

Reg Reps: 10 - 20 - 30 - 40 - 50 - 60 - 70 - 80 - 90 - 100 - 110 - 120 - 130
140 - 150 - 160 - 170 - 180 - 190 - 200 - 210 - 220 - 230 - 240 - 250 - 260
270 - 280 - 290 - 300 - 310 - 320 - 330 - 340 - 350 - 360 - 370 - 380 - 390
400 - 410 - 420 - 430 - 440 - 450 - 460 - 470 - 480 - 490 - 500 - 510 - 520
530 - 540 - 550 - 560 - 570 - 580 - 590 - 600 - 610 - 620 - 630 - 640 - 650
660 - 670 - 680 - 690 - 700 - 710 - 720 - 730 - 740 - 750 - 760 - 770 - 780
790 - 800 - 810 - 820 - 830 - 840 - 850 - 860 - 870 - 880 - 890 - 900 - 910
920 - 930 - 940 - 950 - 960 - 970 - 980 - 990 - 1000

Notes: ____

____

____

____

Reg Reps: 10 - 20 - 30 - 40 - 50 - 60 - 70 - 80 - 90 - 100 - 110 - 120 - 130
140 - 150 - 160 - 170 - 180 - 190 - 200 - 210 - 220 - 230 - 240 - 250 - 260
270 - 280 - 290 - 300 - 310 - 320 - 330 - 340 - 350 - 360 - 370 - 380 - 390
400 - 410 - 420 - 430 - 440 - 450 - 460 - 470 - 480 - 490 - 500 - 510 - 520
530 - 540 - 550 - 560 - 570 - 580 - 590 - 600 - 610 - 620 - 630 - 640 - 650
660 - 670 - 680 - 690 - 700 - 710 - 720 - 730 - 740 - 750 - 760 - 770 - 780
790 - 800 - 810 - 820 - 830 - 840 - 850 - 860 - 870 - 880 - 890 - 900 - 910
920 - 930 - 940 - 950 - 960 - 970 - 980 - 990 - 1000

Notes: ____

____

____

____

Reg Reps: 10 - 20 - 30 - 40 - 50 - 60 - 70 - 80 - 90 - 100 - 110 - 120 - 130
140 - 150 - 160 - 170 - 180 - 190 - 200 - 210 - 220 - 230 - 240 - 250 - 260
270 - 280 - 290 - 300 - 310 - 320 - 330 - 340 - 350 - 360 - 370 - 380 - 390
400 - 410 - 420 - 430 - 440 - 450 - 460 - 470 - 480 - 490 - 500 - 510 - 520
530 - 540 - 550 - 560 - 570 - 580 - 590 - 600 - 610 - 620 - 630 - 640 - 650
660 - 670 - 680 - 690 - 700 - 710 - 720 - 730 - 740 - 750 - 760 - 770 - 780
790 - 800 - 810 - 820 - 830 - 840 - 850 - 860 - 870 - 880 - 890 - 900 - 910
920 - 930 - 940 - 950 - 960 - 970 - 980 - 990 - 1000

Notes: ____

____

____

______________________________________ **End Date:** __________

Defense Against:______________________________ **Start Date:** ____________

Notes:____

____

____

____

Slow Reps: 10 - 20 - 30 - 40 - 50 - 60 - 70 - 80 - 90 - 100 - 110 - 120 - 130
140 - 150 - 160 - 170 - 180 - 190 - 200 - 210 - 220 - 230 - 240 - 250 - 260
270 - 280 - 290 - 300 - 310 - 320 - 330 - 340 - 350 - 360 - 370 - 380 - 390
400 - 410 - 420 - 430 - 440 - 450 - 460 - 470 - 480 - 490 - 500 - 510 - 520
530 - 540 - 550 - 560 - 570 - 580 - 590 - 600 - 610 - 620 - 630 - 640 - 650
660 - 670 - 680 - 690 - 700 - 710 - 720 - 730 - 740 - 750 - 760 - 770 - 780
790 - 800 - 810 - 820 - 830 - 840 - 850 - 860 - 870 - 880 - 890 - 900 - 910
920 - 930 - 940 - 950 - 960 - 970 - 980 - 990 - 1000

Reg Reps: 10 - 20 - 30 - 40 - 50 - 60 - 70 - 80 - 90 - 100 - 110 - 120 - 130
140 - 150 - 160 - 170 - 180 - 190 - 200 - 210 - 220 - 230 - 240 - 250 - 260
270 - 280 - 290 - 300 - 310 - 320 - 330 - 340 - 350 - 360 - 370 - 380 - 390
400 - 410 - 420 - 430 - 440 - 450 - 460 - 470 - 480 - 490 - 500 - 510 - 520
530 - 540 - 550 - 560 - 570 - 580 - 590 - 600 - 610 - 620 - 630 - 640 - 650
660 - 670 - 680 - 690 - 700 - 710 - 720 - 730 - 740 - 750 - 760 - 770 - 780
790 - 800 - 810 - 820 - 830 - 840 - 850 - 860 - 870 - 880 - 890 - 900 - 910
920 - 930 - 940 - 950 - 960 - 970 - 980 - 990 - 1000

Notes: ____

____

____

____

Reg Reps: 10 - 20 - 30 - 40 - 50 - 60 - 70 - 80 - 90 - 100 - 110 - 120 - 130
140 - 150 - 160 - 170 - 180 - 190 - 200 - 210 - 220 - 230 - 240 - 250 - 260
270 - 280 - 290 - 300 - 310 - 320 - 330 - 340 - 350 - 360 - 370 - 380 - 390
400 - 410 - 420 - 430 - 440 - 450 - 460 - 470 - 480 - 490 - 500 - 510 - 520
530 - 540 - 550 - 560 - 570 - 580 - 590 - 600 - 610 - 620 - 630 - 640 - 650
660 - 670 - 680 - 690 - 700 - 710 - 720 - 730 - 740 - 750 - 760 - 770 - 780
790 - 800 - 810 - 820 - 830 - 840 - 850 - 860 - 870 - 880 - 890 - 900 - 910
920 - 930 - 940 - 950 - 960 - 970 - 980 - 990 - 1000

Notes: ____

____

____

____

Reg Reps: 10 - 20 - 30 - 40 - 50 - 60 - 70 - 80 - 90 - 100 - 110 - 120 - 130
140 - 150 - 160 - 170 - 180 - 190 - 200 - 210 - 220 - 230 - 240 - 250 - 260
270 - 280 - 290 - 300 - 310 - 320 - 330 - 340 - 350 - 360 - 370 - 380 - 390
400 - 410 - 420 - 430 - 440 - 450 - 460 - 470 - 480 - 490 - 500 - 510 - 520
530 - 540 - 550 - 560 - 570 - 580 - 590 - 600 - 610 - 620 - 630 - 640 - 650
660 - 670 - 680 - 690 - 700 - 710 - 720 - 730 - 740 - 750 - 760 - 770 - 780
790 - 800 - 810 - 820 - 830 - 840 - 850 - 860 - 870 - 880 - 890 - 900 - 910
920 - 930 - 940 - 950 - 960 - 970 - 980 - 990 - 1000

Notes: ______________________________

____

____

____

Reg Reps: 10 - 20 - 30 - 40 - 50 - 60 - 70 - 80 - 90 - 100 - 110 - 120 - 130
140 - 150 - 160 - 170 - 180 - 190 - 200 - 210 - 220 - 230 - 240 - 250 - 260
270 - 280 - 290 - 300 - 310 - 320 - 330 - 340 - 350 - 360 - 370 - 380 - 390
400 - 410 - 420 - 430 - 440 - 450 - 460 - 470 - 480 - 490 - 500 - 510 - 520
530 - 540 - 550 - 560 - 570 - 580 - 590 - 600 - 610 - 620 - 630 - 640 - 650
660 - 670 - 680 - 690 - 700 - 710 - 720 - 730 - 740 - 750 - 760 - 770 - 780
790 - 800 - 810 - 820 - 830 - 840 - 850 - 860 - 870 - 880 - 890 - 900 - 910
920 - 930 - 940 - 950 - 960 - 970 - 980 - 990 - 1000

Notes: ______________________________

____

____

____

Reg Reps: 10 - 20 - 30 - 40 - 50 - 60 - 70 - 80 - 90 - 100 - 110 - 120 - 130
140 - 150 - 160 - 170 - 180 - 190 - 200 - 210 - 220 - 230 - 240 - 250 - 260
270 - 280 - 290 - 300 - 310 - 320 - 330 - 340 - 350 - 360 - 370 - 380 - 390
400 - 410 - 420 - 430 - 440 - 450 - 460 - 470 - 480 - 490 - 500 - 510 - 520
530 - 540 - 550 - 560 - 570 - 580 - 590 - 600 - 610 - 620 - 630 - 640 - 650
660 - 670 - 680 - 690 - 700 - 710 - 720 - 730 - 740 - 750 - 760 - 770 - 780
790 - 800 - 810 - 820 - 830 - 840 - 850 - 860 - 870 - 880 - 890 - 900 - 910
920 - 930 - 940 - 950 - 960 - 970 - 980 - 990 - 1000

Notes: ______________________________

____

____

______________________________ **End Date:** ____________

Defense Against:_________________________ **Start Date:** __________

Notes:____

Slow Reps: 10 - 20 - 30 - 40 - 50 - 60 - 70 - 80 - 90 - 100 - 110 - 120 - 130
140 - 150 - 160 - 170 - 180 - 190 - 200 - 210 - 220 - 230 - 240 - 250 - 260
270 - 280 - 290 - 300 - 310 - 320 - 330 - 340 - 350 - 360 - 370 - 380 - 390
400 - 410 - 420 - 430 - 440 - 450 - 460 - 470 - 480 - 490 - 500 - 510 - 520
530 - 540 - 550 - 560 - 570 - 580 - 590 - 600 - 610 - 620 - 630 - 640 - 650
660 - 670 - 680 - 690 - 700 - 710 - 720 - 730 - 740 - 750 - 760 - 770 - 780
790 - 800 - 810 - 820 - 830 - 840 - 850 - 860 - 870 - 880 - 890 - 900 - 910
920 - 930 - 940 - 950 - 960 - 970 - 980 - 990 - 1000

Reg Reps: 10 - 20 - 30 - 40 - 50 - 60 - 70 - 80 - 90 - 100 - 110 - 120 - 130
140 - 150 - 160 - 170 - 180 - 190 - 200 - 210 - 220 - 230 - 240 - 250 - 260
270 - 280 - 290 - 300 - 310 - 320 - 330 - 340 - 350 - 360 - 370 - 380 - 390
400 - 410 - 420 - 430 - 440 - 450 - 460 - 470 - 480 - 490 - 500 - 510 - 520
530 - 540 - 550 - 560 - 570 - 580 - 590 - 600 - 610 - 620 - 630 - 640 - 650
660 - 670 - 680 - 690 - 700 - 710 - 720 - 730 - 740 - 750 - 760 - 770 - 780
790 - 800 - 810 - 820 - 830 - 840 - 850 - 860 - 870 - 880 - 890 - 900 - 910
920 - 930 - 940 - 950 - 960 - 970 - 980 - 990 - 1000

Notes: ____

Reg Reps: 10 - 20 - 30 - 40 - 50 - 60 - 70 - 80 - 90 - 100 - 110 - 120 - 130
140 - 150 - 160 - 170 - 180 - 190 - 200 - 210 - 220 - 230 - 240 - 250 - 260
270 - 280 - 290 - 300 - 310 - 320 - 330 - 340 - 350 - 360 - 370 - 380 - 390
400 - 410 - 420 - 430 - 440 - 450 - 460 - 470 - 480 - 490 - 500 - 510 - 520
530 - 540 - 550 - 560 - 570 - 580 - 590 - 600 - 610 - 620 - 630 - 640 - 650
660 - 670 - 680 - 690 - 700 - 710 - 720 - 730 - 740 - 750 - 760 - 770 - 780
790 - 800 - 810 - 820 - 830 - 840 - 850 - 860 - 870 - 880 - 890 - 900 - 910
920 - 930 - 940 - 950 - 960 - 970 - 980 - 990 - 1000

Notes: ____

Reg Reps: 10 - 20 - 30 - 40 - 50 - 60 - 70 - 80 - 90 - 100 - 110 - 120 - 130 140 - 150 - 160 - 170 - 180 - 190 - 200 - 210 - 220 - 230 - 240 - 250 - 260 270 - 280 - 290 - 300 - 310 - 320 - 330 - 340 - 350 - 360 - 370 - 380 - 390 400 - 410 - 420 - 430 - 440 - 450 - 460 - 470 - 480 - 490 - 500 - 510 - 520 530 - 540 - 550 - 560 - 570 - 580 - 590 - 600 - 610 - 620 - 630 - 640 - 650 660 - 670 - 680 - 690 - 700 - 710 - 720 - 730 - 740 - 750 - 760 - 770 - 780 790 - 800 - 810 - 820 - 830 - 840 - 850 - 860 - 870 - 880 - 890 - 900 - 910 920 - 930 - 940 - 950 - 960 - 970 - 980 - 990 - 1000

Notes: ______________________________

____

____

____

Reg Reps: 10 - 20 - 30 - 40 - 50 - 60 - 70 - 80 - 90 - 100 - 110 - 120 - 130 140 - 150 - 160 - 170 - 180 - 190 - 200 - 210 - 220 - 230 - 240 - 250 - 260 270 - 280 - 290 - 300 - 310 - 320 - 330 - 340 - 350 - 360 - 370 - 380 - 390 400 - 410 - 420 - 430 - 440 - 450 - 460 - 470 - 480 - 490 - 500 - 510 - 520 530 - 540 - 550 - 560 - 570 - 580 - 590 - 600 - 610 - 620 - 630 - 640 - 650 660 - 670 - 680 - 690 - 700 - 710 - 720 - 730 - 740 - 750 - 760 - 770 - 780 790 - 800 - 810 - 820 - 830 - 840 - 850 - 860 - 870 - 880 - 890 - 900 - 910 920 - 930 - 940 - 950 - 960 - 970 - 980 - 990 - 1000

Notes: ______________________________

____

____

____

Reg Reps: 10 - 20 - 30 - 40 - 50 - 60 - 70 - 80 - 90 - 100 - 110 - 120 - 130 140 - 150 - 160 - 170 - 180 - 190 - 200 - 210 - 220 - 230 - 240 - 250 - 260 270 - 280 - 290 - 300 - 310 - 320 - 330 - 340 - 350 - 360 - 370 - 380 - 390 400 - 410 - 420 - 430 - 440 - 450 - 460 - 470 - 480 - 490 - 500 - 510 - 520 530 - 540 - 550 - 560 - 570 - 580 - 590 - 600 - 610 - 620 - 630 - 640 - 650 660 - 670 - 680 - 690 - 700 - 710 - 720 - 730 - 740 - 750 - 760 - 770 - 780 790 - 800 - 810 - 820 - 830 - 840 - 850 - 860 - 870 - 880 - 890 - 900 - 910 920 - 930 - 940 - 950 - 960 - 970 - 980 - 990 - 1000

Notes: ______________________________

____

____

______________________________ **End Date:** ____________

Defense Against:__________________________ **Start Date:** __________

Notes:____

Slow Reps: 10 - 20 - 30 - 40 - 50 - 60 - 70 - 80 - 90 - 100 - 110 - 120 - 130
140 - 150 - 160 - 170 - 180 - 190 - 200 - 210 - 220 - 230 - 240 - 250 - 260
270 - 280 - 290 - 300 - 310 - 320 - 330 - 340 - 350 - 360 - 370 - 380 - 390
400 - 410 - 420 - 430 - 440 - 450 - 460 - 470 - 480 - 490 - 500 - 510 - 520
530 - 540 - 550 - 560 - 570 - 580 - 590 - 600 - 610 - 620 - 630 - 640 - 650
660 - 670 - 680 - 690 - 700 - 710 - 720 - 730 - 740 - 750 - 760 - 770 - 780
790 - 800 - 810 - 820 - 830 - 840 - 850 - 860 - 870 - 880 - 890 - 900 - 910
920 - 930 - 940 - 950 - 960 - 970 - 980 - 990 - 1000

Reg Reps: 10 - 20 - 30 - 40 - 50 - 60 - 70 - 80 - 90 - 100 - 110 - 120 - 130
140 - 150 - 160 - 170 - 180 - 190 - 200 - 210 - 220 - 230 - 240 - 250 - 260
270 - 280 - 290 - 300 - 310 - 320 - 330 - 340 - 350 - 360 - 370 - 380 - 390
400 - 410 - 420 - 430 - 440 - 450 - 460 - 470 - 480 - 490 - 500 - 510 - 520
530 - 540 - 550 - 560 - 570 - 580 - 590 - 600 - 610 - 620 - 630 - 640 - 650
660 - 670 - 680 - 690 - 700 - 710 - 720 - 730 - 740 - 750 - 760 - 770 - 780
790 - 800 - 810 - 820 - 830 - 840 - 850 - 860 - 870 - 880 - 890 - 900 - 910
920 - 930 - 940 - 950 - 960 - 970 - 980 - 990 - 1000

Notes: ____

Reg Reps: 10 - 20 - 30 - 40 - 50 - 60 - 70 - 80 - 90 - 100 - 110 - 120 - 130
140 - 150 - 160 - 170 - 180 - 190 - 200 - 210 - 220 - 230 - 240 - 250 - 260
270 - 280 - 290 - 300 - 310 - 320 - 330 - 340 - 350 - 360 - 370 - 380 - 390
400 - 410 - 420 - 430 - 440 - 450 - 460 - 470 - 480 - 490 - 500 - 510 - 520
530 - 540 - 550 - 560 - 570 - 580 - 590 - 600 - 610 - 620 - 630 - 640 - 650
660 - 670 - 680 - 690 - 700 - 710 - 720 - 730 - 740 - 750 - 760 - 770 - 780
790 - 800 - 810 - 820 - 830 - 840 - 850 - 860 - 870 - 880 - 890 - 900 - 910
920 - 930 - 940 - 950 - 960 - 970 - 980 - 990 - 1000

Notes: ____

Reg Reps: 10 - 20 - 30 - 40 - 50 - 60 - 70 - 80 - 90 - 100 - 110 - 120 - 130 - 140 - 150 - 160 - 170 - 180 - 190 - 200 - 210 - 220 - 230 - 240 - 250 - 260 - 270 - 280 - 290 - 300 - 310 - 320 - 330 - 340 - 350 - 360 - 370 - 380 - 390 - 400 - 410 - 420 - 430 - 440 - 450 - 460 - 470 - 480 - 490 - 500 - 510 - 520 - 530 - 540 - 550 - 560 - 570 - 580 - 590 - 600 - 610 - 620 - 630 - 640 - 650 - 660 - 670 - 680 - 690 - 700 - 710 - 720 - 730 - 740 - 750 - 760 - 770 - 780 - 790 - 800 - 810 - 820 - 830 - 840 - 850 - 860 - 870 - 880 - 890 - 900 - 910 - 920 - 930 - 940 - 950 - 960 - 970 - 980 - 990 - 1000

Notes: ______________________________

____

____

____

Reg Reps: 10 - 20 - 30 - 40 - 50 - 60 - 70 - 80 - 90 - 100 - 110 - 120 - 130 - 140 - 150 - 160 - 170 - 180 - 190 - 200 - 210 - 220 - 230 - 240 - 250 - 260 - 270 - 280 - 290 - 300 - 310 - 320 - 330 - 340 - 350 - 360 - 370 - 380 - 390 - 400 - 410 - 420 - 430 - 440 - 450 - 460 - 470 - 480 - 490 - 500 - 510 - 520 - 530 - 540 - 550 - 560 - 570 - 580 - 590 - 600 - 610 - 620 - 630 - 640 - 650 - 660 - 670 - 680 - 690 - 700 - 710 - 720 - 730 - 740 - 750 - 760 - 770 - 780 - 790 - 800 - 810 - 820 - 830 - 840 - 850 - 860 - 870 - 880 - 890 - 900 - 910 - 920 - 930 - 940 - 950 - 960 - 970 - 980 - 990 - 1000

Notes: ______________________________

____

____

____

Reg Reps: 10 - 20 - 30 - 40 - 50 - 60 - 70 - 80 - 90 - 100 - 110 - 120 - 130 - 140 - 150 - 160 - 170 - 180 - 190 - 200 - 210 - 220 - 230 - 240 - 250 - 260 - 270 - 280 - 290 - 300 - 310 - 320 - 330 - 340 - 350 - 360 - 370 - 380 - 390 - 400 - 410 - 420 - 430 - 440 - 450 - 460 - 470 - 480 - 490 - 500 - 510 - 520 - 530 - 540 - 550 - 560 - 570 - 580 - 590 - 600 - 610 - 620 - 630 - 640 - 650 - 660 - 670 - 680 - 690 - 700 - 710 - 720 - 730 - 740 - 750 - 760 - 770 - 780 - 790 - 800 - 810 - 820 - 830 - 840 - 850 - 860 - 870 - 880 - 890 - 900 - 910 - 920 - 930 - 940 - 950 - 960 - 970 - 980 - 990 - 1000

Notes: ______________________________

____

____

______________________________ **End Date:** ____________

Defense Against:____________________________ **Start Date:** __________

Notes:____

Slow Reps: 10 - 20 - 30 - 40 - 50 - 60 - 70 - 80 - 90 - 100 - 110 - 120 - 130
140 - 150 - 160 - 170 - 180 - 190 - 200 - 210 - 220 - 230 - 240 - 250 - 260
270 - 280 - 290 - 300 - 310 - 320 - 330 - 340 - 350 - 360 - 370 - 380 - 390
400 - 410 - 420 - 430 - 440 - 450 - 460 - 470 - 480 - 490 - 500 - 510 - 520
530 - 540 - 550 - 560 - 570 - 580 - 590 - 600 - 610 - 620 - 630 - 640 - 650
660 - 670 - 680 - 690 - 700 - 710 - 720 - 730 - 740 - 750 - 760 - 770 - 780
790 - 800 - 810 - 820 - 830 - 840 - 850 - 860 - 870 - 880 - 890 - 900 - 910
920 - 930 - 940 - 950 - 960 - 970 - 980 - 990 - 1000

Reg Reps: 10 - 20 - 30 - 40 - 50 - 60 - 70 - 80 - 90 - 100 - 110 - 120 - 130
140 - 150 - 160 - 170 - 180 - 190 - 200 - 210 - 220 - 230 - 240 - 250 - 260
270 - 280 - 290 - 300 - 310 - 320 - 330 - 340 - 350 - 360 - 370 - 380 - 390
400 - 410 - 420 - 430 - 440 - 450 - 460 - 470 - 480 - 490 - 500 - 510 - 520
530 - 540 - 550 - 560 - 570 - 580 - 590 - 600 - 610 - 620 - 630 - 640 - 650
660 - 670 - 680 - 690 - 700 - 710 - 720 - 730 - 740 - 750 - 760 - 770 - 780
790 - 800 - 810 - 820 - 830 - 840 - 850 - 860 - 870 - 880 - 890 - 900 - 910
920 - 930 - 940 - 950 - 960 - 970 - 980 - 990 - 1000

Notes: ____

Reg Reps: 10 - 20 - 30 - 40 - 50 - 60 - 70 - 80 - 90 - 100 - 110 - 120 - 130
140 - 150 - 160 - 170 - 180 - 190 - 200 - 210 - 220 - 230 - 240 - 250 - 260
270 - 280 - 290 - 300 - 310 - 320 - 330 - 340 - 350 - 360 - 370 - 380 - 390
400 - 410 - 420 - 430 - 440 - 450 - 460 - 470 - 480 - 490 - 500 - 510 - 520
530 - 540 - 550 - 560 - 570 - 580 - 590 - 600 - 610 - 620 - 630 - 640 - 650
660 - 670 - 680 - 690 - 700 - 710 - 720 - 730 - 740 - 750 - 760 - 770 - 780
790 - 800 - 810 - 820 - 830 - 840 - 850 - 860 - 870 - 880 - 890 - 900 - 910
920 - 930 - 940 - 950 - 960 - 970 - 980 - 990 - 1000

Notes: ____

Reg Reps: 10 - 20 - 30 - 40 - 50 - 60 - 70 - 80 - 90 - 100 - 110 - 120 - 130 140 - 150 - 160 - 170 - 180 - 190 - 200 - 210 - 220 - 230 - 240 - 250 - 260 270 - 280 - 290 - 300 - 310 - 320 - 330 - 340 - 350 - 360 - 370 - 380 - 390 400 - 410 - 420 - 430 - 440 - 450 - 460 - 470 - 480 - 490 - 500 - 510 - 520 530 - 540 - 550 - 560 - 570 - 580 - 590 - 600 - 610 - 620 - 630 - 640 - 650 660 - 670 - 680 - 690 - 700 - 710 - 720 - 730 - 740 - 750 - 760 - 770 - 780 790 - 800 - 810 - 820 - 830 - 840 - 850 - 860 - 870 - 880 - 890 - 900 - 910 920 - 930 - 940 - 950 - 960 - 970 - 980 - 990 - 1000

Notes: ______________________________

Reg Reps: 10 - 20 - 30 - 40 - 50 - 60 - 70 - 80 - 90 - 100 - 110 - 120 - 130 140 - 150 - 160 - 170 - 180 - 190 - 200 - 210 - 220 - 230 - 240 - 250 - 260 270 - 280 - 290 - 300 - 310 - 320 - 330 - 340 - 350 - 360 - 370 - 380 - 390 400 - 410 - 420 - 430 - 440 - 450 - 460 - 470 - 480 - 490 - 500 - 510 - 520 530 - 540 - 550 - 560 - 570 - 580 - 590 - 600 - 610 - 620 - 630 - 640 - 650 660 - 670 - 680 - 690 - 700 - 710 - 720 - 730 - 740 - 750 - 760 - 770 - 780 790 - 800 - 810 - 820 - 830 - 840 - 850 - 860 - 870 - 880 - 890 - 900 - 910 920 - 930 - 940 - 950 - 960 - 970 - 980 - 990 - 1000

Notes: ______________________________

Reg Reps: 10 - 20 - 30 - 40 - 50 - 60 - 70 - 80 - 90 - 100 - 110 - 120 - 130 140 - 150 - 160 - 170 - 180 - 190 - 200 - 210 - 220 - 230 - 240 - 250 - 260 270 - 280 - 290 - 300 - 310 - 320 - 330 - 340 - 350 - 360 - 370 - 380 - 390 400 - 410 - 420 - 430 - 440 - 450 - 460 - 470 - 480 - 490 - 500 - 510 - 520 530 - 540 - 550 - 560 - 570 - 580 - 590 - 600 - 610 - 620 - 630 - 640 - 650 660 - 670 - 680 - 690 - 700 - 710 - 720 - 730 - 740 - 750 - 760 - 770 - 780 790 - 800 - 810 - 820 - 830 - 840 - 850 - 860 - 870 - 880 - 890 - 900 - 910 920 - 930 - 940 - 950 - 960 - 970 - 980 - 990 - 1000

Notes: ______________________________

End Date: ______________

Defense Against:__________________________ **Start Date:** __________

Notes:____

____

____

____

Slow Reps: 10 - 20 - 30 - 40 - 50 - 60 - 70 - 80 - 90 - 100 - 110 - 120 - 130
140 - 150 - 160 - 170 - 180 - 190 - 200 - 210 - 220 - 230 - 240 - 250 - 260
270 - 280 - 290 - 300 - 310 - 320 - 330 - 340 - 350 - 360 - 370 - 380 - 390
400 - 410 - 420 - 430 - 440 - 450 - 460 - 470 - 480 - 490 - 500 - 510 - 520
530 - 540 - 550 - 560 - 570 - 580 - 590 - 600 - 610 - 620 - 630 - 640 - 650
660 - 670 - 680 - 690 - 700 - 710 - 720 - 730 - 740 - 750 - 760 - 770 - 780
790 - 800 - 810 - 820 - 830 - 840 - 850 - 860 - 870 - 880 - 890 - 900 - 910
920 - 930 - 940 - 950 - 960 - 970 - 980 - 990 - 1000

Reg Reps: 10 - 20 - 30 - 40 - 50 - 60 - 70 - 80 - 90 - 100 - 110 - 120 - 130
140 - 150 - 160 - 170 - 180 - 190 - 200 - 210 - 220 - 230 - 240 - 250 - 260
270 - 280 - 290 - 300 - 310 - 320 - 330 - 340 - 350 - 360 - 370 - 380 - 390
400 - 410 - 420 - 430 - 440 - 450 - 460 - 470 - 480 - 490 - 500 - 510 - 520
530 - 540 - 550 - 560 - 570 - 580 - 590 - 600 - 610 - 620 - 630 - 640 - 650
660 - 670 - 680 - 690 - 700 - 710 - 720 - 730 - 740 - 750 - 760 - 770 - 780
790 - 800 - 810 - 820 - 830 - 840 - 850 - 860 - 870 - 880 - 890 - 900 - 910
920 - 930 - 940 - 950 - 960 - 970 - 980 - 990 - 1000

Notes: ____

____

____

____

Reg Reps: 10 - 20 - 30 - 40 - 50 - 60 - 70 - 80 - 90 - 100 - 110 - 120 - 130
140 - 150 - 160 - 170 - 180 - 190 - 200 - 210 - 220 - 230 - 240 - 250 - 260
270 - 280 - 290 - 300 - 310 - 320 - 330 - 340 - 350 - 360 - 370 - 380 - 390
400 - 410 - 420 - 430 - 440 - 450 - 460 - 470 - 480 - 490 - 500 - 510 - 520
530 - 540 - 550 - 560 - 570 - 580 - 590 - 600 - 610 - 620 - 630 - 640 - 650
660 - 670 - 680 - 690 - 700 - 710 - 720 - 730 - 740 - 750 - 760 - 770 - 780
790 - 800 - 810 - 820 - 830 - 840 - 850 - 860 - 870 - 880 - 890 - 900 - 910
920 - 930 - 940 - 950 - 960 - 970 - 980 - 990 - 1000

Notes: ____

____

____

____

Reg Reps: 10 - 20 - 30 - 40 - 50 - 60 - 70 - 80 - 90 - 100 - 110 - 120 - 130
140 - 150 - 160 - 170 - 180 - 190 - 200 - 210 - 220 - 230 - 240 - 250 - 260
270 - 280 - 290 - 300 - 310 - 320 - 330 - 340 - 350 - 360 - 370 - 380 - 390
400 - 410 - 420 - 430 - 440 - 450 - 460 - 470 - 480 - 490 - 500 - 510 - 520
530 - 540 - 550 - 560 - 570 - 580 - 590 - 600 - 610 - 620 - 630 - 640 - 650
660 - 670 - 680 - 690 - 700 - 710 - 720 - 730 - 740 - 750 - 760 - 770 - 780
790 - 800 - 810 - 820 - 830 - 840 - 850 - 860 - 870 - 880 - 890 - 900 - 910
920 - 930 - 940 - 950 - 960 - 970 - 980 - 990 - 1000

Notes: ____

____

____

____

Reg Reps: 10 - 20 - 30 - 40 - 50 - 60 - 70 - 80 - 90 - 100 - 110 - 120 - 130
140 - 150 - 160 - 170 - 180 - 190 - 200 - 210 - 220 - 230 - 240 - 250 - 260
270 - 280 - 290 - 300 - 310 - 320 - 330 - 340 - 350 - 360 - 370 - 380 - 390
400 - 410 - 420 - 430 - 440 - 450 - 460 - 470 - 480 - 490 - 500 - 510 - 520
530 - 540 - 550 - 560 - 570 - 580 - 590 - 600 - 610 - 620 - 630 - 640 - 650
660 - 670 - 680 - 690 - 700 - 710 - 720 - 730 - 740 - 750 - 760 - 770 - 780
790 - 800 - 810 - 820 - 830 - 840 - 850 - 860 - 870 - 880 - 890 - 900 - 910
920 - 930 - 940 - 950 - 960 - 970 - 980 - 990 - 1000

Notes: ____

____

____

____

Reg Reps: 10 - 20 - 30 - 40 - 50 - 60 - 70 - 80 - 90 - 100 - 110 - 120 - 130
140 - 150 - 160 - 170 - 180 - 190 - 200 - 210 - 220 - 230 - 240 - 250 - 260
270 - 280 - 290 - 300 - 310 - 320 - 330 - 340 - 350 - 360 - 370 - 380 - 390
400 - 410 - 420 - 430 - 440 - 450 - 460 - 470 - 480 - 490 - 500 - 510 - 520
530 - 540 - 550 - 560 - 570 - 580 - 590 - 600 - 610 - 620 - 630 - 640 - 650
660 - 670 - 680 - 690 - 700 - 710 - 720 - 730 - 740 - 750 - 760 - 770 - 780
790 - 800 - 810 - 820 - 830 - 840 - 850 - 860 - 870 - 880 - 890 - 900 - 910
920 - 930 - 940 - 950 - 960 - 970 - 980 - 990 - 1000

Notes: ____

____

____

______________________________ **End Date:** ______________

Defense Against:____________________________ **Start Date:** ___________

Notes:____

____

____

____

Slow Reps: 10 - 20 - 30 - 40 - 50 - 60 - 70 - 80 - 90 - 100 - 110 - 120 - 130
140 - 150 - 160 - 170 - 180 - 190 - 200 - 210 - 220 - 230 - 240 - 250 - 260
270 - 280 - 290 - 300 - 310 - 320 - 330 - 340 - 350 - 360 - 370 - 380 - 390
400 - 410 - 420 - 430 - 440 - 450 - 460 - 470 - 480 - 490 - 500 - 510 - 520
530 - 540 - 550 - 560 - 570 - 580 - 590 - 600 - 610 - 620 - 630 - 640 - 650
660 - 670 - 680 - 690 - 700 - 710 - 720 - 730 - 740 - 750 - 760 - 770 - 780
790 - 800 - 810 - 820 - 830 - 840 - 850 - 860 - 870 - 880 - 890 - 900 - 910
920 - 930 - 940 - 950 - 960 - 970 - 980 - 990 - 1000

Reg Reps: 10 - 20 - 30 - 40 - 50 - 60 - 70 - 80 - 90 - 100 - 110 - 120 - 130
140 - 150 - 160 - 170 - 180 - 190 - 200 - 210 - 220 - 230 - 240 - 250 - 260
270 - 280 - 290 - 300 - 310 - 320 - 330 - 340 - 350 - 360 - 370 - 380 - 390
400 - 410 - 420 - 430 - 440 - 450 - 460 - 470 - 480 - 490 - 500 - 510 - 520
530 - 540 - 550 - 560 - 570 - 580 - 590 - 600 - 610 - 620 - 630 - 640 - 650
660 - 670 - 680 - 690 - 700 - 710 - 720 - 730 - 740 - 750 - 760 - 770 - 780
790 - 800 - 810 - 820 - 830 - 840 - 850 - 860 - 870 - 880 - 890 - 900 - 910
920 - 930 - 940 - 950 - 960 - 970 - 980 - 990 - 1000

Notes: ____

____

____

____

Reg Reps: 10 - 20 - 30 - 40 - 50 - 60 - 70 - 80 - 90 - 100 - 110 - 120 - 130
140 - 150 - 160 - 170 - 180 - 190 - 200 - 210 - 220 - 230 - 240 - 250 - 260
270 - 280 - 290 - 300 - 310 - 320 - 330 - 340 - 350 - 360 - 370 - 380 - 390
400 - 410 - 420 - 430 - 440 - 450 - 460 - 470 - 480 - 490 - 500 - 510 - 520
530 - 540 - 550 - 560 - 570 - 580 - 590 - 600 - 610 - 620 - 630 - 640 - 650
660 - 670 - 680 - 690 - 700 - 710 - 720 - 730 - 740 - 750 - 760 - 770 - 780
790 - 800 - 810 - 820 - 830 - 840 - 850 - 860 - 870 - 880 - 890 - 900 - 910
920 - 930 - 940 - 950 - 960 - 970 - 980 - 990 - 1000

Notes: ____

____

____

____

Reg Reps: 10 - 20 - 30 - 40 - 50 - 60 - 70 - 80 - 90 - 100 - 110 - 120 - 130
140 - 150 - 160 - 170 - 180 - 190 - 200 - 210 - 220 - 230 - 240 - 250 - 260
270 - 280 - 290 - 300 - 310 - 320 - 330 - 340 - 350 - 360 - 370 - 380 - 390
400 - 410 - 420 - 430 - 440 - 450 - 460 - 470 - 480 - 490 - 500 - 510 - 520
530 - 540 - 550 - 560 - 570 - 580 - 590 - 600 - 610 - 620 - 630 - 640 - 650
660 - 670 - 680 - 690 - 700 - 710 - 720 - 730 - 740 - 750 - 760 - 770 - 780
790 - 800 - 810 - 820 - 830 - 840 - 850 - 860 - 870 - 880 - 890 - 900 - 910
920 - 930 - 940 - 950 - 960 - 970 - 980 - 990 - 1000

Notes: ____

____

____

____

Reg Reps: 10 - 20 - 30 - 40 - 50 - 60 - 70 - 80 - 90 - 100 - 110 - 120 - 130
140 - 150 - 160 - 170 - 180 - 190 - 200 - 210 - 220 - 230 - 240 - 250 - 260
270 - 280 - 290 - 300 - 310 - 320 - 330 - 340 - 350 - 360 - 370 - 380 - 390
400 - 410 - 420 - 430 - 440 - 450 - 460 - 470 - 480 - 490 - 500 - 510 - 520
530 - 540 - 550 - 560 - 570 - 580 - 590 - 600 - 610 - 620 - 630 - 640 - 650
660 - 670 - 680 - 690 - 700 - 710 - 720 - 730 - 740 - 750 - 760 - 770 - 780
790 - 800 - 810 - 820 - 830 - 840 - 850 - 860 - 870 - 880 - 890 - 900 - 910
920 - 930 - 940 - 950 - 960 - 970 - 980 - 990 - 1000

Notes: ____

____

____

____

Reg Reps: 10 - 20 - 30 - 40 - 50 - 60 - 70 - 80 - 90 - 100 - 110 - 120 - 130
140 - 150 - 160 - 170 - 180 - 190 - 200 - 210 - 220 - 230 - 240 - 250 - 260
270 - 280 - 290 - 300 - 310 - 320 - 330 - 340 - 350 - 360 - 370 - 380 - 390
400 - 410 - 420 - 430 - 440 - 450 - 460 - 470 - 480 - 490 - 500 - 510 - 520
530 - 540 - 550 - 560 - 570 - 580 - 590 - 600 - 610 - 620 - 630 - 640 - 650
660 - 670 - 680 - 690 - 700 - 710 - 720 - 730 - 740 - 750 - 760 - 770 - 780
790 - 800 - 810 - 820 - 830 - 840 - 850 - 860 - 870 - 880 - 890 - 900 - 910
920 - 930 - 940 - 950 - 960 - 970 - 980 - 990 - 1000

Notes: ____

____

____

______________________________ **End Date:** __________

Defense Against:_____________________________ **Start Date:** __________
Notes:____

____

____

____

Slow Reps: 10 - 20 - 30 - 40 - 50 - 60 - 70 - 80 - 90 - 100 - 110 - 120 - 130
140 - 150 - 160 - 170 - 180 - 190 - 200 - 210 - 220 - 230 - 240 - 250 - 260
270 - 280 - 290 - 300 - 310 - 320 - 330 - 340 - 350 - 360 - 370 - 380 - 390
400 - 410 - 420 - 430 - 440 - 450 - 460 - 470 - 480 - 490 - 500 - 510 - 520
530 - 540 - 550 - 560 - 570 - 580 - 590 - 600 - 610 - 620 - 630 - 640 - 650
660 - 670 - 680 - 690 - 700 - 710 - 720 - 730 - 740 - 750 - 760 - 770 - 780
790 - 800 - 810 - 820 - 830 - 840 - 850 - 860 - 870 - 880 - 890 - 900 - 910
920 - 930 - 940 - 950 - 960 - 970 - 980 - 990 - 1000

Reg Reps: 10 - 20 - 30 - 40 - 50 - 60 - 70 - 80 - 90 - 100 - 110 - 120 - 130
140 - 150 - 160 - 170 - 180 - 190 - 200 - 210 - 220 - 230 - 240 - 250 - 260
270 - 280 - 290 - 300 - 310 - 320 - 330 - 340 - 350 - 360 - 370 - 380 - 390
400 - 410 - 420 - 430 - 440 - 450 - 460 - 470 - 480 - 490 - 500 - 510 - 520
530 - 540 - 550 - 560 - 570 - 580 - 590 - 600 - 610 - 620 - 630 - 640 - 650
660 - 670 - 680 - 690 - 700 - 710 - 720 - 730 - 740 - 750 - 760 - 770 - 780
790 - 800 - 810 - 820 - 830 - 840 - 850 - 860 - 870 - 880 - 890 - 900 - 910
920 - 930 - 940 - 950 - 960 - 970 - 980 - 990 - 1000
Notes: ____

____

____

____

Reg Reps: 10 - 20 - 30 - 40 - 50 - 60 - 70 - 80 - 90 - 100 - 110 - 120 - 130
140 - 150 - 160 - 170 - 180 - 190 - 200 - 210 - 220 - 230 - 240 - 250 - 260
270 - 280 - 290 - 300 - 310 - 320 - 330 - 340 - 350 - 360 - 370 - 380 - 390
400 - 410 - 420 - 430 - 440 - 450 - 460 - 470 - 480 - 490 - 500 - 510 - 520
530 - 540 - 550 - 560 - 570 - 580 - 590 - 600 - 610 - 620 - 630 - 640 - 650
660 - 670 - 680 - 690 - 700 - 710 - 720 - 730 - 740 - 750 - 760 - 770 - 780
790 - 800 - 810 - 820 - 830 - 840 - 850 - 860 - 870 - 880 - 890 - 900 - 910
920 - 930 - 940 - 950 - 960 - 970 - 980 - 990 - 1000
Notes: ____

____

____

____

Reg Reps: 10 - 20 - 30 - 40 - 50 - 60 - 70 - 80 - 90 - 100 - 110 - 120 - 130 140 - 150 - 160 - 170 - 180 - 190 - 200 - 210 - 220 - 230 - 240 - 250 - 260 270 - 280 - 290 - 300 - 310 - 320 - 330 - 340 - 350 - 360 - 370 - 380 - 390 400 - 410 - 420 - 430 - 440 - 450 - 460 - 470 - 480 - 490 - 500 - 510 - 520 530 - 540 - 550 - 560 - 570 - 580 - 590 - 600 - 610 - 620 - 630 - 640 - 650 660 - 670 - 680 - 690 - 700 - 710 - 720 - 730 - 740 - 750 - 760 - 770 - 780 790 - 800 - 810 - 820 - 830 - 840 - 850 - 860 - 870 - 880 - 890 - 900 - 910 920 - 930 - 940 - 950 - 960 - 970 - 980 - 990 - 1000

Notes: ______________________________

____

____

____

Reg Reps: 10 - 20 - 30 - 40 - 50 - 60 - 70 - 80 - 90 - 100 - 110 - 120 - 130 140 - 150 - 160 - 170 - 180 - 190 - 200 - 210 - 220 - 230 - 240 - 250 - 260 270 - 280 - 290 - 300 - 310 - 320 - 330 - 340 - 350 - 360 - 370 - 380 - 390 400 - 410 - 420 - 430 - 440 - 450 - 460 - 470 - 480 - 490 - 500 - 510 - 520 530 - 540 - 550 - 560 - 570 - 580 - 590 - 600 - 610 - 620 - 630 - 640 - 650 660 - 670 - 680 - 690 - 700 - 710 - 720 - 730 - 740 - 750 - 760 - 770 - 780 790 - 800 - 810 - 820 - 830 - 840 - 850 - 860 - 870 - 880 - 890 - 900 - 910 920 - 930 - 940 - 950 - 960 - 970 - 980 - 990 - 1000

Notes: ______________________________

____

____

____

Reg Reps: 10 - 20 - 30 - 40 - 50 - 60 - 70 - 80 - 90 - 100 - 110 - 120 - 130 140 - 150 - 160 - 170 - 180 - 190 - 200 - 210 - 220 - 230 - 240 - 250 - 260 270 - 280 - 290 - 300 - 310 - 320 - 330 - 340 - 350 - 360 - 370 - 380 - 390 400 - 410 - 420 - 430 - 440 - 450 - 460 - 470 - 480 - 490 - 500 - 510 - 520 530 - 540 - 550 - 560 - 570 - 580 - 590 - 600 - 610 - 620 - 630 - 640 - 650 660 - 670 - 680 - 690 - 700 - 710 - 720 - 730 - 740 - 750 - 760 - 770 - 780 790 - 800 - 810 - 820 - 830 - 840 - 850 - 860 - 870 - 880 - 890 - 900 - 910 920 - 930 - 940 - 950 - 960 - 970 - 980 - 990 - 1000

Notes: ______________________________

____

____

______________________________ **End Date:** __________

Defense Against:___________________________ **Start Date:** ___________

Notes:____

Slow Reps: 10 - 20 - 30 - 40 - 50 - 60 - 70 - 80 - 90 - 100 - 110 - 120 - 130 - 140 - 150 - 160 - 170 - 180 - 190 - 200 - 210 - 220 - 230 - 240 - 250 - 260 - 270 - 280 - 290 - 300 - 310 - 320 - 330 - 340 - 350 - 360 - 370 - 380 - 390 - 400 - 410 - 420 - 430 - 440 - 450 - 460 - 470 - 480 - 490 - 500 - 510 - 520 - 530 - 540 - 550 - 560 - 570 - 580 - 590 - 600 - 610 - 620 - 630 - 640 - 650 - 660 - 670 - 680 - 690 - 700 - 710 - 720 - 730 - 740 - 750 - 760 - 770 - 780 - 790 - 800 - 810 - 820 - 830 - 840 - 850 - 860 - 870 - 880 - 890 - 900 - 910 - 920 - 930 - 940 - 950 - 960 - 970 - 980 - 990 - 1000

Reg Reps: 10 - 20 - 30 - 40 - 50 - 60 - 70 - 80 - 90 - 100 - 110 - 120 - 130 - 140 - 150 - 160 - 170 - 180 - 190 - 200 - 210 - 220 - 230 - 240 - 250 - 260 - 270 - 280 - 290 - 300 - 310 - 320 - 330 - 340 - 350 - 360 - 370 - 380 - 390 - 400 - 410 - 420 - 430 - 440 - 450 - 460 - 470 - 480 - 490 - 500 - 510 - 520 - 530 - 540 - 550 - 560 - 570 - 580 - 590 - 600 - 610 - 620 - 630 - 640 - 650 - 660 - 670 - 680 - 690 - 700 - 710 - 720 - 730 - 740 - 750 - 760 - 770 - 780 - 790 - 800 - 810 - 820 - 830 - 840 - 850 - 860 - 870 - 880 - 890 - 900 - 910 - 920 - 930 - 940 - 950 - 960 - 970 - 980 - 990 - 1000

Notes: ____

Reg Reps: 10 - 20 - 30 - 40 - 50 - 60 - 70 - 80 - 90 - 100 - 110 - 120 - 130 - 140 - 150 - 160 - 170 - 180 - 190 - 200 - 210 - 220 - 230 - 240 - 250 - 260 - 270 - 280 - 290 - 300 - 310 - 320 - 330 - 340 - 350 - 360 - 370 - 380 - 390 - 400 - 410 - 420 - 430 - 440 - 450 - 460 - 470 - 480 - 490 - 500 - 510 - 520 - 530 - 540 - 550 - 560 - 570 - 580 - 590 - 600 - 610 - 620 - 630 - 640 - 650 - 660 - 670 - 680 - 690 - 700 - 710 - 720 - 730 - 740 - 750 - 760 - 770 - 780 - 790 - 800 - 810 - 820 - 830 - 840 - 850 - 860 - 870 - 880 - 890 - 900 - 910 - 920 - 930 - 940 - 950 - 960 - 970 - 980 - 990 - 1000

Notes: ____

Reg Reps: 10 - 20 - 30 - 40 - 50 - 60 - 70 - 80 - 90 - 100 - 110 - 120 - 130
140 - 150 - 160 - 170 - 180 - 190 - 200 - 210 - 220 - 230 - 240 - 250 - 260
270 - 280 - 290 - 300 - 310 - 320 - 330 - 340 - 350 - 360 - 370 - 380 - 390
400 - 410 - 420 - 430 - 440 - 450 - 460 - 470 - 480 - 490 - 500 - 510 - 520
530 - 540 - 550 - 560 - 570 - 580 - 590 - 600 - 610 - 620 - 630 - 640 - 650
660 - 670 - 680 - 690 - 700 - 710 - 720 - 730 - 740 - 750 - 760 - 770 - 780
790 - 800 - 810 - 820 - 830 - 840 - 850 - 860 - 870 - 880 - 890 - 900 - 910
920 - 930 - 940 - 950 - 960 - 970 - 980 - 990 - 1000

Notes: ______________________________

____

____

____

Reg Reps: 10 - 20 - 30 - 40 - 50 - 60 - 70 - 80 - 90 - 100 - 110 - 120 - 130
140 - 150 - 160 - 170 - 180 - 190 - 200 - 210 - 220 - 230 - 240 - 250 - 260
270 - 280 - 290 - 300 - 310 - 320 - 330 - 340 - 350 - 360 - 370 - 380 - 390
400 - 410 - 420 - 430 - 440 - 450 - 460 - 470 - 480 - 490 - 500 - 510 - 520
530 - 540 - 550 - 560 - 570 - 580 - 590 - 600 - 610 - 620 - 630 - 640 - 650
660 - 670 - 680 - 690 - 700 - 710 - 720 - 730 - 740 - 750 - 760 - 770 - 780
790 - 800 - 810 - 820 - 830 - 840 - 850 - 860 - 870 - 880 - 890 - 900 - 910
920 - 930 - 940 - 950 - 960 - 970 - 980 - 990 - 1000

Notes: ______________________________

____

____

____

Reg Reps: 10 - 20 - 30 - 40 - 50 - 60 - 70 - 80 - 90 - 100 - 110 - 120 - 130
140 - 150 - 160 - 170 - 180 - 190 - 200 - 210 - 220 - 230 - 240 - 250 - 260
270 - 280 - 290 - 300 - 310 - 320 - 330 - 340 - 350 - 360 - 370 - 380 - 390
400 - 410 - 420 - 430 - 440 - 450 - 460 - 470 - 480 - 490 - 500 - 510 - 520
530 - 540 - 550 - 560 - 570 - 580 - 590 - 600 - 610 - 620 - 630 - 640 - 650
660 - 670 - 680 - 690 - 700 - 710 - 720 - 730 - 740 - 750 - 760 - 770 - 780
790 - 800 - 810 - 820 - 830 - 840 - 850 - 860 - 870 - 880 - 890 - 900 - 910
920 - 930 - 940 - 950 - 960 - 970 - 980 - 990 - 1000

Notes: ______________________________

____

____

______________________________ **End Date:** __________

Defense Against:________________________ **Start Date:** __________
Notes:____

Slow Reps: 10 - 20 - 30 - 40 - 50 - 60 - 70 - 80 - 90 - 100 - 110 - 120 - 130
140 - 150 - 160 - 170 - 180 - 190 - 200 - 210 - 220 - 230 - 240 - 250 - 260
270 - 280 - 290 - 300 - 310 - 320 - 330 - 340 - 350 - 360 - 370 - 380 - 390
400 - 410 - 420 - 430 - 440 - 450 - 460 - 470 - 480 - 490 - 500 - 510 - 520
530 - 540 - 550 - 560 - 570 - 580 - 590 - 600 - 610 - 620 - 630 - 640 - 650
660 - 670 - 680 - 690 - 700 - 710 - 720 - 730 - 740 - 750 - 760 - 770 - 780
790 - 800 - 810 - 820 - 830 - 840 - 850 - 860 - 870 - 880 - 890 - 900 - 910
920 - 930 - 940 - 950 - 960 - 970 - 980 - 990 - 1000

Reg Reps: 10 - 20 - 30 - 40 - 50 - 60 - 70 - 80 - 90 - 100 - 110 - 120 - 130
140 - 150 - 160 - 170 - 180 - 190 - 200 - 210 - 220 - 230 - 240 - 250 - 260
270 - 280 - 290 - 300 - 310 - 320 - 330 - 340 - 350 - 360 - 370 - 380 - 390
400 - 410 - 420 - 430 - 440 - 450 - 460 - 470 - 480 - 490 - 500 - 510 - 520
530 - 540 - 550 - 560 - 570 - 580 - 590 - 600 - 610 - 620 - 630 - 640 - 650
660 - 670 - 680 - 690 - 700 - 710 - 720 - 730 - 740 - 750 - 760 - 770 - 780
790 - 800 - 810 - 820 - 830 - 840 - 850 - 860 - 870 - 880 - 890 - 900 - 910
920 - 930 - 940 - 950 - 960 - 970 - 980 - 990 - 1000
Notes: ____

Reg Reps: 10 - 20 - 30 - 40 - 50 - 60 - 70 - 80 - 90 - 100 - 110 - 120 - 130
140 - 150 - 160 - 170 - 180 - 190 - 200 - 210 - 220 - 230 - 240 - 250 - 260
270 - 280 - 290 - 300 - 310 - 320 - 330 - 340 - 350 - 360 - 370 - 380 - 390
400 - 410 - 420 - 430 - 440 - 450 - 460 - 470 - 480 - 490 - 500 - 510 - 520
530 - 540 - 550 - 560 - 570 - 580 - 590 - 600 - 610 - 620 - 630 - 640 - 650
660 - 670 - 680 - 690 - 700 - 710 - 720 - 730 - 740 - 750 - 760 - 770 - 780
790 - 800 - 810 - 820 - 830 - 840 - 850 - 860 - 870 - 880 - 890 - 900 - 910
920 - 930 - 940 - 950 - 960 - 970 - 980 - 990 - 1000
Notes: ____

Reg Reps: 10 - 20 - 30 - 40 - 50 - 60 - 70 - 80 - 90 - 100 - 110 - 120 - 130
140 - 150 - 160 - 170 - 180 - 190 - 200 - 210 - 220 - 230 - 240 - 250 - 260
270 - 280 - 290 - 300 - 310 - 320 - 330 - 340 - 350 - 360 - 370 - 380 - 390
400 - 410 - 420 - 430 - 440 - 450 - 460 - 470 - 480 - 490 - 500 - 510 - 520
530 - 540 - 550 - 560 - 570 - 580 - 590 - 600 - 610 - 620 - 630 - 640 - 650
660 - 670 - 680 - 690 - 700 - 710 - 720 - 730 - 740 - 750 - 760 - 770 - 780
790 - 800 - 810 - 820 - 830 - 840 - 850 - 860 - 870 - 880 - 890 - 900 - 910
920 - 930 - 940 - 950 - 960 - 970 - 980 - 990 - 1000

Notes: ______________________________

____

____

____

Reg Reps: 10 - 20 - 30 - 40 - 50 - 60 - 70 - 80 - 90 - 100 - 110 - 120 - 130
140 - 150 - 160 - 170 - 180 - 190 - 200 - 210 - 220 - 230 - 240 - 250 - 260
270 - 280 - 290 - 300 - 310 - 320 - 330 - 340 - 350 - 360 - 370 - 380 - 390
400 - 410 - 420 - 430 - 440 - 450 - 460 - 470 - 480 - 490 - 500 - 510 - 520
530 - 540 - 550 - 560 - 570 - 580 - 590 - 600 - 610 - 620 - 630 - 640 - 650
660 - 670 - 680 - 690 - 700 - 710 - 720 - 730 - 740 - 750 - 760 - 770 - 780
790 - 800 - 810 - 820 - 830 - 840 - 850 - 860 - 870 - 880 - 890 - 900 - 910
920 - 930 - 940 - 950 - 960 - 970 - 980 - 990 - 1000

Notes: ______________________________

____

____

____

Reg Reps: 10 - 20 - 30 - 40 - 50 - 60 - 70 - 80 - 90 - 100 - 110 - 120 - 130
140 - 150 - 160 - 170 - 180 - 190 - 200 - 210 - 220 - 230 - 240 - 250 - 260
270 - 280 - 290 - 300 - 310 - 320 - 330 - 340 - 350 - 360 - 370 - 380 - 390
400 - 410 - 420 - 430 - 440 - 450 - 460 - 470 - 480 - 490 - 500 - 510 - 520
530 - 540 - 550 - 560 - 570 - 580 - 590 - 600 - 610 - 620 - 630 - 640 - 650
660 - 670 - 680 - 690 - 700 - 710 - 720 - 730 - 740 - 750 - 760 - 770 - 780
790 - 800 - 810 - 820 - 830 - 840 - 850 - 860 - 870 - 880 - 890 - 900 - 910
920 - 930 - 940 - 950 - 960 - 970 - 980 - 990 - 1000

Notes: ______________________________

____

____

______________________________ **End Date:** ______________

Defense Against:________________________ **Start Date:** __________

Notes:____

____

____

____

Slow Reps: 10 - 20 - 30 - 40 - 50 - 60 - 70 - 80 - 90 - 100 - 110 - 120 - 130
140 - 150 - 160 - 170 - 180 - 190 - 200 - 210 - 220 - 230 - 240 - 250 - 260
270 - 280 - 290 - 300 - 310 - 320 - 330 - 340 - 350 - 360 - 370 - 380 - 390
400 - 410 - 420 - 430 - 440 - 450 - 460 - 470 - 480 - 490 - 500 - 510 - 520
530 - 540 - 550 - 560 - 570 - 580 - 590 - 600 - 610 - 620 - 630 - 640 - 650
660 - 670 - 680 - 690 - 700 - 710 - 720 - 730 - 740 - 750 - 760 - 770 - 780
790 - 800 - 810 - 820 - 830 - 840 - 850 - 860 - 870 - 880 - 890 - 900 - 910
920 - 930 - 940 - 950 - 960 - 970 - 980 - 990 - 1000

Reg Reps: 10 - 20 - 30 - 40 - 50 - 60 - 70 - 80 - 90 - 100 - 110 - 120 - 130
140 - 150 - 160 - 170 - 180 - 190 - 200 - 210 - 220 - 230 - 240 - 250 - 260
270 - 280 - 290 - 300 - 310 - 320 - 330 - 340 - 350 - 360 - 370 - 380 - 390
400 - 410 - 420 - 430 - 440 - 450 - 460 - 470 - 480 - 490 - 500 - 510 - 520
530 - 540 - 550 - 560 - 570 - 580 - 590 - 600 - 610 - 620 - 630 - 640 - 650
660 - 670 - 680 - 690 - 700 - 710 - 720 - 730 - 740 - 750 - 760 - 770 - 780
790 - 800 - 810 - 820 - 830 - 840 - 850 - 860 - 870 - 880 - 890 - 900 - 910
920 - 930 - 940 - 950 - 960 - 970 - 980 - 990 - 1000

Notes: ____

____

____

____

Reg Reps: 10 - 20 - 30 - 40 - 50 - 60 - 70 - 80 - 90 - 100 - 110 - 120 - 130
140 - 150 - 160 - 170 - 180 - 190 - 200 - 210 - 220 - 230 - 240 - 250 - 260
270 - 280 - 290 - 300 - 310 - 320 - 330 - 340 - 350 - 360 - 370 - 380 - 390
400 - 410 - 420 - 430 - 440 - 450 - 460 - 470 - 480 - 490 - 500 - 510 - 520
530 - 540 - 550 - 560 - 570 - 580 - 590 - 600 - 610 - 620 - 630 - 640 - 650
660 - 670 - 680 - 690 - 700 - 710 - 720 - 730 - 740 - 750 - 760 - 770 - 780
790 - 800 - 810 - 820 - 830 - 840 - 850 - 860 - 870 - 880 - 890 - 900 - 910
920 - 930 - 940 - 950 - 960 - 970 - 980 - 990 - 1000

Notes: ____

____

____

____

Reg Reps: 10 - 20 - 30 - 40 - 50 - 60 - 70 - 80 - 90 - 100 - 110 - 120 - 130
140 - 150 - 160 - 170 - 180 - 190 - 200 - 210 - 220 - 230 - 240 - 250 - 260
270 - 280 - 290 - 300 - 310 - 320 - 330 - 340 - 350 - 360 - 370 - 380 - 390
400 - 410 - 420 - 430 - 440 - 450 - 460 - 470 - 480 - 490 - 500 - 510 - 520
530 - 540 - 550 - 560 - 570 - 580 - 590 - 600 - 610 - 620 - 630 - 640 - 650
660 - 670 - 680 - 690 - 700 - 710 - 720 - 730 - 740 - 750 - 760 - 770 - 780
790 - 800 - 810 - 820 - 830 - 840 - 850 - 860 - 870 - 880 - 890 - 900 - 910
920 - 930 - 940 - 950 - 960 - 970 - 980 - 990 - 1000

Notes: ____

____

____

____

Reg Reps: 10 - 20 - 30 - 40 - 50 - 60 - 70 - 80 - 90 - 100 - 110 - 120 - 130
140 - 150 - 160 - 170 - 180 - 190 - 200 - 210 - 220 - 230 - 240 - 250 - 260
270 - 280 - 290 - 300 - 310 - 320 - 330 - 340 - 350 - 360 - 370 - 380 - 390
400 - 410 - 420 - 430 - 440 - 450 - 460 - 470 - 480 - 490 - 500 - 510 - 520
530 - 540 - 550 - 560 - 570 - 580 - 590 - 600 - 610 - 620 - 630 - 640 - 650
660 - 670 - 680 - 690 - 700 - 710 - 720 - 730 - 740 - 750 - 760 - 770 - 780
790 - 800 - 810 - 820 - 830 - 840 - 850 - 860 - 870 - 880 - 890 - 900 - 910
920 - 930 - 940 - 950 - 960 - 970 - 980 - 990 - 1000

Notes: ____

____

____

____

Reg Reps: 10 - 20 - 30 - 40 - 50 - 60 - 70 - 80 - 90 - 100 - 110 - 120 - 130
140 - 150 - 160 - 170 - 180 - 190 - 200 - 210 - 220 - 230 - 240 - 250 - 260
270 - 280 - 290 - 300 - 310 - 320 - 330 - 340 - 350 - 360 - 370 - 380 - 390
400 - 410 - 420 - 430 - 440 - 450 - 460 - 470 - 480 - 490 - 500 - 510 - 520
530 - 540 - 550 - 560 - 570 - 580 - 590 - 600 - 610 - 620 - 630 - 640 - 650
660 - 670 - 680 - 690 - 700 - 710 - 720 - 730 - 740 - 750 - 760 - 770 - 780
790 - 800 - 810 - 820 - 830 - 840 - 850 - 860 - 870 - 880 - 890 - 900 - 910
920 - 930 - 940 - 950 - 960 - 970 - 980 - 990 - 1000

Notes: ____

____

____

________________________________ **End Date:** __________

Defense Against:__________________________ **Start Date:** __________
Notes:____

Slow Reps: 10 - 20 - 30 - 40 - 50 - 60 - 70 - 80 - 90 - 100 - 110 - 120 - 130
140 - 150 - 160 - 170 - 180 - 190 - 200 - 210 - 220 - 230 - 240 - 250 - 260
270 - 280 - 290 - 300 - 310 - 320 - 330 - 340 - 350 - 360 - 370 - 380 - 390
400 - 410 - 420 - 430 - 440 - 450 - 460 - 470 - 480 - 490 - 500 - 510 - 520
530 - 540 - 550 - 560 - 570 - 580 - 590 - 600 - 610 - 620 - 630 - 640 - 650
660 - 670 - 680 - 690 - 700 - 710 - 720 - 730 - 740 - 750 - 760 - 770 - 780
790 - 800 - 810 - 820 - 830 - 840 - 850 - 860 - 870 - 880 - 890 - 900 - 910
920 - 930 - 940 - 950 - 960 - 970 - 980 - 990 - 1000

Reg Reps: 10 - 20 - 30 - 40 - 50 - 60 - 70 - 80 - 90 - 100 - 110 - 120 - 130
140 - 150 - 160 - 170 - 180 - 190 - 200 - 210 - 220 - 230 - 240 - 250 - 260
270 - 280 - 290 - 300 - 310 - 320 - 330 - 340 - 350 - 360 - 370 - 380 - 390
400 - 410 - 420 - 430 - 440 - 450 - 460 - 470 - 480 - 490 - 500 - 510 - 520
530 - 540 - 550 - 560 - 570 - 580 - 590 - 600 - 610 - 620 - 630 - 640 - 650
660 - 670 - 680 - 690 - 700 - 710 - 720 - 730 - 740 - 750 - 760 - 770 - 780
790 - 800 - 810 - 820 - 830 - 840 - 850 - 860 - 870 - 880 - 890 - 900 - 910
920 - 930 - 940 - 950 - 960 - 970 - 980 - 990 - 1000
Notes: ____

Reg Reps: 10 - 20 - 30 - 40 - 50 - 60 - 70 - 80 - 90 - 100 - 110 - 120 - 130
140 - 150 - 160 - 170 - 180 - 190 - 200 - 210 - 220 - 230 - 240 - 250 - 260
270 - 280 - 290 - 300 - 310 - 320 - 330 - 340 - 350 - 360 - 370 - 380 - 390
400 - 410 - 420 - 430 - 440 - 450 - 460 - 470 - 480 - 490 - 500 - 510 - 520
530 - 540 - 550 - 560 - 570 - 580 - 590 - 600 - 610 - 620 - 630 - 640 - 650
660 - 670 - 680 - 690 - 700 - 710 - 720 - 730 - 740 - 750 - 760 - 770 - 780
790 - 800 - 810 - 820 - 830 - 840 - 850 - 860 - 870 - 880 - 890 - 900 - 910
920 - 930 - 940 - 950 - 960 - 970 - 980 - 990 - 1000
Notes: ____

Reg Reps: 10 - 20 - 30 - 40 - 50 - 60 - 70 - 80 - 90 - 100 - 110 - 120 - 130
140 - 150 - 160 - 170 - 180 - 190 - 200 - 210 - 220 - 230 - 240 - 250 - 260
270 - 280 - 290 - 300 - 310 - 320 - 330 - 340 - 350 - 360 - 370 - 380 - 390
400 - 410 - 420 - 430 - 440 - 450 - 460 - 470 - 480 - 490 - 500 - 510 - 520
530 - 540 - 550 - 560 - 570 - 580 - 590 - 600 - 610 - 620 - 630 - 640 - 650
660 - 670 - 680 - 690 - 700 - 710 - 720 - 730 - 740 - 750 - 760 - 770 - 780
790 - 800 - 810 - 820 - 830 - 840 - 850 - 860 - 870 - 880 - 890 - 900 - 910
920 - 930 - 940 - 950 - 960 - 970 - 980 - 990 - 1000

Notes: ____

____

____

____

Reg Reps: 10 - 20 - 30 - 40 - 50 - 60 - 70 - 80 - 90 - 100 - 110 - 120 - 130
140 - 150 - 160 - 170 - 180 - 190 - 200 - 210 - 220 - 230 - 240 - 250 - 260
270 - 280 - 290 - 300 - 310 - 320 - 330 - 340 - 350 - 360 - 370 - 380 - 390
400 - 410 - 420 - 430 - 440 - 450 - 460 - 470 - 480 - 490 - 500 - 510 - 520
530 - 540 - 550 - 560 - 570 - 580 - 590 - 600 - 610 - 620 - 630 - 640 - 650
660 - 670 - 680 - 690 - 700 - 710 - 720 - 730 - 740 - 750 - 760 - 770 - 780
790 - 800 - 810 - 820 - 830 - 840 - 850 - 860 - 870 - 880 - 890 - 900 - 910
920 - 930 - 940 - 950 - 960 - 970 - 980 - 990 - 1000

Notes: ____

____

____

____

Reg Reps: 10 - 20 - 30 - 40 - 50 - 60 - 70 - 80 - 90 - 100 - 110 - 120 - 130
140 - 150 - 160 - 170 - 180 - 190 - 200 - 210 - 220 - 230 - 240 - 250 - 260
270 - 280 - 290 - 300 - 310 - 320 - 330 - 340 - 350 - 360 - 370 - 380 - 390
400 - 410 - 420 - 430 - 440 - 450 - 460 - 470 - 480 - 490 - 500 - 510 - 520
530 - 540 - 550 - 560 - 570 - 580 - 590 - 600 - 610 - 620 - 630 - 640 - 650
660 - 670 - 680 - 690 - 700 - 710 - 720 - 730 - 740 - 750 - 760 - 770 - 780
790 - 800 - 810 - 820 - 830 - 840 - 850 - 860 - 870 - 880 - 890 - 900 - 910
920 - 930 - 940 - 950 - 960 - 970 - 980 - 990 - 1000

Notes: ____

____

____

______________________________ **End Date:** ____________

Defense Against: ______________________________ **Start Date:** __________

Notes: ____

____

____

____

Slow Reps: 10 - 20 - 30 - 40 - 50 - 60 - 70 - 80 - 90 - 100 - 110 - 120 - 130
140 - 150 - 160 - 170 - 180 - 190 - 200 - 210 - 220 - 230 - 240 - 250 - 260
270 - 280 - 290 - 300 - 310 - 320 - 330 - 340 - 350 - 360 - 370 - 380 - 390
400 - 410 - 420 - 430 - 440 - 450 - 460 - 470 - 480 - 490 - 500 - 510 - 520
530 - 540 - 550 - 560 - 570 - 580 - 590 - 600 - 610 - 620 - 630 - 640 - 650
660 - 670 - 680 - 690 - 700 - 710 - 720 - 730 - 740 - 750 - 760 - 770 - 780
790 - 800 - 810 - 820 - 830 - 840 - 850 - 860 - 870 - 880 - 890 - 900 - 910
920 - 930 - 940 - 950 - 960 - 970 - 980 - 990 - 1000

Reg Reps: 10 - 20 - 30 - 40 - 50 - 60 - 70 - 80 - 90 - 100 - 110 - 120 - 130
140 - 150 - 160 - 170 - 180 - 190 - 200 - 210 - 220 - 230 - 240 - 250 - 260
270 - 280 - 290 - 300 - 310 - 320 - 330 - 340 - 350 - 360 - 370 - 380 - 390
400 - 410 - 420 - 430 - 440 - 450 - 460 - 470 - 480 - 490 - 500 - 510 - 520
530 - 540 - 550 - 560 - 570 - 580 - 590 - 600 - 610 - 620 - 630 - 640 - 650
660 - 670 - 680 - 690 - 700 - 710 - 720 - 730 - 740 - 750 - 760 - 770 - 780
790 - 800 - 810 - 820 - 830 - 840 - 850 - 860 - 870 - 880 - 890 - 900 - 910
920 - 930 - 940 - 950 - 960 - 970 - 980 - 990 - 1000

Notes: ____

____

____

____

Reg Reps: 10 - 20 - 30 - 40 - 50 - 60 - 70 - 80 - 90 - 100 - 110 - 120 - 130
140 - 150 - 160 - 170 - 180 - 190 - 200 - 210 - 220 - 230 - 240 - 250 - 260
270 - 280 - 290 - 300 - 310 - 320 - 330 - 340 - 350 - 360 - 370 - 380 - 390
400 - 410 - 420 - 430 - 440 - 450 - 460 - 470 - 480 - 490 - 500 - 510 - 520
530 - 540 - 550 - 560 - 570 - 580 - 590 - 600 - 610 - 620 - 630 - 640 - 650
660 - 670 - 680 - 690 - 700 - 710 - 720 - 730 - 740 - 750 - 760 - 770 - 780
790 - 800 - 810 - 820 - 830 - 840 - 850 - 860 - 870 - 880 - 890 - 900 - 910
920 - 930 - 940 - 950 - 960 - 970 - 980 - 990 - 1000

Notes: ____

____

____

____

Reg Reps: 10 - 20 - 30 - 40 - 50 - 60 - 70 - 80 - 90 - 100 - 110 - 120 - 130
140 - 150 - 160 - 170 - 180 - 190 - 200 - 210 - 220 - 230 - 240 - 250 - 260
270 - 280 - 290 - 300 - 310 - 320 - 330 - 340 - 350 - 360 - 370 - 380 - 390
400 - 410 - 420 - 430 - 440 - 450 - 460 - 470 - 480 - 490 - 500 - 510 - 520
530 - 540 - 550 - 560 - 570 - 580 - 590 - 600 - 610 - 620 - 630 - 640 - 650
660 - 670 - 680 - 690 - 700 - 710 - 720 - 730 - 740 - 750 - 760 - 770 - 780
790 - 800 - 810 - 820 - 830 - 840 - 850 - 860 - 870 - 880 - 890 - 900 - 910
920 - 930 - 940 - 950 - 960 - 970 - 980 - 990 - 1000

Notes: ____________________

____

____

____

Reg Reps: 10 - 20 - 30 - 40 - 50 - 60 - 70 - 80 - 90 - 100 - 110 - 120 - 130
140 - 150 - 160 - 170 - 180 - 190 - 200 - 210 - 220 - 230 - 240 - 250 - 260
270 - 280 - 290 - 300 - 310 - 320 - 330 - 340 - 350 - 360 - 370 - 380 - 390
400 - 410 - 420 - 430 - 440 - 450 - 460 - 470 - 480 - 490 - 500 - 510 - 520
530 - 540 - 550 - 560 - 570 - 580 - 590 - 600 - 610 - 620 - 630 - 640 - 650
660 - 670 - 680 - 690 - 700 - 710 - 720 - 730 - 740 - 750 - 760 - 770 - 780
790 - 800 - 810 - 820 - 830 - 840 - 850 - 860 - 870 - 880 - 890 - 900 - 910
920 - 930 - 940 - 950 - 960 - 970 - 980 - 990 - 1000

Notes: ____________________

____

____

____

Reg Reps: 10 - 20 - 30 - 40 - 50 - 60 - 70 - 80 - 90 - 100 - 110 - 120 - 130
140 - 150 - 160 - 170 - 180 - 190 - 200 - 210 - 220 - 230 - 240 - 250 - 260
270 - 280 - 290 - 300 - 310 - 320 - 330 - 340 - 350 - 360 - 370 - 380 - 390
400 - 410 - 420 - 430 - 440 - 450 - 460 - 470 - 480 - 490 - 500 - 510 - 520
530 - 540 - 550 - 560 - 570 - 580 - 590 - 600 - 610 - 620 - 630 - 640 - 650
660 - 670 - 680 - 690 - 700 - 710 - 720 - 730 - 740 - 750 - 760 - 770 - 780
790 - 800 - 810 - 820 - 830 - 840 - 850 - 860 - 870 - 880 - 890 - 900 - 910
920 - 930 - 940 - 950 - 960 - 970 - 980 - 990 - 1000

Notes: ____________________

____

____

____________________ **End Date:** __________

Defense Against:__________________________ **Start Date:** __________

Notes:____

Slow Reps: 10 - 20 - 30 - 40 - 50 - 60 - 70 - 80 - 90 - 100 - 110 - 120 - 130
140 - 150 - 160 - 170 - 180 - 190 - 200 - 210 - 220 - 230 - 240 - 250 - 260
270 - 280 - 290 - 300 - 310 - 320 - 330 - 340 - 350 - 360 - 370 - 380 - 390
400 - 410 - 420 - 430 - 440 - 450 - 460 - 470 - 480 - 490 - 500 - 510 - 520
530 - 540 - 550 - 560 - 570 - 580 - 590 - 600 - 610 - 620 - 630 - 640 - 650
660 - 670 - 680 - 690 - 700 - 710 - 720 - 730 - 740 - 750 - 760 - 770 - 780
790 - 800 - 810 - 820 - 830 - 840 - 850 - 860 - 870 - 880 - 890 - 900 - 910
920 - 930 - 940 - 950 - 960 - 970 - 980 - 990 - 1000

Reg Reps: 10 - 20 - 30 - 40 - 50 - 60 - 70 - 80 - 90 - 100 - 110 - 120 - 130
140 - 150 - 160 - 170 - 180 - 190 - 200 - 210 - 220 - 230 - 240 - 250 - 260
270 - 280 - 290 - 300 - 310 - 320 - 330 - 340 - 350 - 360 - 370 - 380 - 390
400 - 410 - 420 - 430 - 440 - 450 - 460 - 470 - 480 - 490 - 500 - 510 - 520
530 - 540 - 550 - 560 - 570 - 580 - 590 - 600 - 610 - 620 - 630 - 640 - 650
660 - 670 - 680 - 690 - 700 - 710 - 720 - 730 - 740 - 750 - 760 - 770 - 780
790 - 800 - 810 - 820 - 830 - 840 - 850 - 860 - 870 - 880 - 890 - 900 - 910
920 - 930 - 940 - 950 - 960 - 970 - 980 - 990 - 1000

Notes: ____

Reg Reps: 10 - 20 - 30 - 40 - 50 - 60 - 70 - 80 - 90 - 100 - 110 - 120 - 130
140 - 150 - 160 - 170 - 180 - 190 - 200 - 210 - 220 - 230 - 240 - 250 - 260
270 - 280 - 290 - 300 - 310 - 320 - 330 - 340 - 350 - 360 - 370 - 380 - 390
400 - 410 - 420 - 430 - 440 - 450 - 460 - 470 - 480 - 490 - 500 - 510 - 520
530 - 540 - 550 - 560 - 570 - 580 - 590 - 600 - 610 - 620 - 630 - 640 - 650
660 - 670 - 680 - 690 - 700 - 710 - 720 - 730 - 740 - 750 - 760 - 770 - 780
790 - 800 - 810 - 820 - 830 - 840 - 850 - 860 - 870 - 880 - 890 - 900 - 910
920 - 930 - 940 - 950 - 960 - 970 - 980 - 990 - 1000

Notes: ____

Reg Reps: 10 - 20 - 30 - 40 - 50 - 60 - 70 - 80 - 90 - 100 - 110 - 120 - 130
140 - 150 - 160 - 170 - 180 - 190 - 200 - 210 - 220 - 230 - 240 - 250 - 260
270 - 280 - 290 - 300 - 310 - 320 - 330 - 340 - 350 - 360 - 370 - 380 - 390
400 - 410 - 420 - 430 - 440 - 450 - 460 - 470 - 480 - 490 - 500 - 510 - 520
530 - 540 - 550 - 560 - 570 - 580 - 590 - 600 - 610 - 620 - 630 - 640 - 650
660 - 670 - 680 - 690 - 700 - 710 - 720 - 730 - 740 - 750 - 760 - 770 - 780
790 - 800 - 810 - 820 - 830 - 840 - 850 - 860 - 870 - 880 - 890 - 900 - 910
920 - 930 - 940 - 950 - 960 - 970 - 980 - 990 - 1000

Notes: ____

____

____

____

Reg Reps: 10 - 20 - 30 - 40 - 50 - 60 - 70 - 80 - 90 - 100 - 110 - 120 - 130
140 - 150 - 160 - 170 - 180 - 190 - 200 - 210 - 220 - 230 - 240 - 250 - 260
270 - 280 - 290 - 300 - 310 - 320 - 330 - 340 - 350 - 360 - 370 - 380 - 390
400 - 410 - 420 - 430 - 440 - 450 - 460 - 470 - 480 - 490 - 500 - 510 - 520
530 - 540 - 550 - 560 - 570 - 580 - 590 - 600 - 610 - 620 - 630 - 640 - 650
660 - 670 - 680 - 690 - 700 - 710 - 720 - 730 - 740 - 750 - 760 - 770 - 780
790 - 800 - 810 - 820 - 830 - 840 - 850 - 860 - 870 - 880 - 890 - 900 - 910
920 - 930 - 940 - 950 - 960 - 970 - 980 - 990 - 1000

Notes: ____

____

____

____

Reg Reps: 10 - 20 - 30 - 40 - 50 - 60 - 70 - 80 - 90 - 100 - 110 - 120 - 130
140 - 150 - 160 - 170 - 180 - 190 - 200 - 210 - 220 - 230 - 240 - 250 - 260
270 - 280 - 290 - 300 - 310 - 320 - 330 - 340 - 350 - 360 - 370 - 380 - 390
400 - 410 - 420 - 430 - 440 - 450 - 460 - 470 - 480 - 490 - 500 - 510 - 520
530 - 540 - 550 - 560 - 570 - 580 - 590 - 600 - 610 - 620 - 630 - 640 - 650
660 - 670 - 680 - 690 - 700 - 710 - 720 - 730 - 740 - 750 - 760 - 770 - 780
790 - 800 - 810 - 820 - 830 - 840 - 850 - 860 - 870 - 880 - 890 - 900 - 910
920 - 930 - 940 - 950 - 960 - 970 - 980 - 990 - 1000

Notes: ____

____

____

______________________________ **End Date:** ____________

Defense Against:_________________________ **Start Date:** __________
Notes:____

Slow Reps: 10 - 20 - 30 - 40 - 50 - 60 - 70 - 80 - 90 - 100 - 110 - 120 - 130 - 140 - 150 - 160 - 170 - 180 - 190 - 200 - 210 - 220 - 230 - 240 - 250 - 260 - 270 - 280 - 290 - 300 - 310 - 320 - 330 - 340 - 350 - 360 - 370 - 380 - 390 - 400 - 410 - 420 - 430 - 440 - 450 - 460 - 470 - 480 - 490 - 500 - 510 - 520 - 530 - 540 - 550 - 560 - 570 - 580 - 590 - 600 - 610 - 620 - 630 - 640 - 650 - 660 - 670 - 680 - 690 - 700 - 710 - 720 - 730 - 740 - 750 - 760 - 770 - 780 - 790 - 800 - 810 - 820 - 830 - 840 - 850 - 860 - 870 - 880 - 890 - 900 - 910 - 920 - 930 - 940 - 950 - 960 - 970 - 980 - 990 - 1000

Reg Reps: 10 - 20 - 30 - 40 - 50 - 60 - 70 - 80 - 90 - 100 - 110 - 120 - 130 - 140 - 150 - 160 - 170 - 180 - 190 - 200 - 210 - 220 - 230 - 240 - 250 - 260 - 270 - 280 - 290 - 300 - 310 - 320 - 330 - 340 - 350 - 360 - 370 - 380 - 390 - 400 - 410 - 420 - 430 - 440 - 450 - 460 - 470 - 480 - 490 - 500 - 510 - 520 - 530 - 540 - 550 - 560 - 570 - 580 - 590 - 600 - 610 - 620 - 630 - 640 - 650 - 660 - 670 - 680 - 690 - 700 - 710 - 720 - 730 - 740 - 750 - 760 - 770 - 780 - 790 - 800 - 810 - 820 - 830 - 840 - 850 - 860 - 870 - 880 - 890 - 900 - 910 - 920 - 930 - 940 - 950 - 960 - 970 - 980 - 990 - 1000

Notes: ____

Reg Reps: 10 - 20 - 30 - 40 - 50 - 60 - 70 - 80 - 90 - 100 - 110 - 120 - 130 - 140 - 150 - 160 - 170 - 180 - 190 - 200 - 210 - 220 - 230 - 240 - 250 - 260 - 270 - 280 - 290 - 300 - 310 - 320 - 330 - 340 - 350 - 360 - 370 - 380 - 390 - 400 - 410 - 420 - 430 - 440 - 450 - 460 - 470 - 480 - 490 - 500 - 510 - 520 - 530 - 540 - 550 - 560 - 570 - 580 - 590 - 600 - 610 - 620 - 630 - 640 - 650 - 660 - 670 - 680 - 690 - 700 - 710 - 720 - 730 - 740 - 750 - 760 - 770 - 780 - 790 - 800 - 810 - 820 - 830 - 840 - 850 - 860 - 870 - 880 - 890 - 900 - 910 - 920 - 930 - 940 - 950 - 960 - 970 - 980 - 990 - 1000

Notes: ____

Reg Reps: 10 - 20 - 30 - 40 - 50 - 60 - 70 - 80 - 90 - 100 - 110 - 120 - 130
140 - 150 - 160 - 170 - 180 - 190 - 200 - 210 - 220 - 230 - 240 - 250 - 260
270 - 280 - 290 - 300 - 310 - 320 - 330 - 340 - 350 - 360 - 370 - 380 - 390
400 - 410 - 420 - 430 - 440 - 450 - 460 - 470 - 480 - 490 - 500 - 510 - 520
530 - 540 - 550 - 560 - 570 - 580 - 590 - 600 - 610 - 620 - 630 - 640 - 650
660 - 670 - 680 - 690 - 700 - 710 - 720 - 730 - 740 - 750 - 760 - 770 - 780
790 - 800 - 810 - 820 - 830 - 840 - 850 - 860 - 870 - 880 - 890 - 900 - 910
920 - 930 - 940 - 950 - 960 - 970 - 980 - 990 - 1000

Notes: ______________________________

____

____

____

Reg Reps: 10 - 20 - 30 - 40 - 50 - 60 - 70 - 80 - 90 - 100 - 110 - 120 - 130
140 - 150 - 160 - 170 - 180 - 190 - 200 - 210 - 220 - 230 - 240 - 250 - 260
270 - 280 - 290 - 300 - 310 - 320 - 330 - 340 - 350 - 360 - 370 - 380 - 390
400 - 410 - 420 - 430 - 440 - 450 - 460 - 470 - 480 - 490 - 500 - 510 - 520
530 - 540 - 550 - 560 - 570 - 580 - 590 - 600 - 610 - 620 - 630 - 640 - 650
660 - 670 - 680 - 690 - 700 - 710 - 720 - 730 - 740 - 750 - 760 - 770 - 780
790 - 800 - 810 - 820 - 830 - 840 - 850 - 860 - 870 - 880 - 890 - 900 - 910
920 - 930 - 940 - 950 - 960 - 970 - 980 - 990 - 1000

Notes: ______________________________

____

____

____

Reg Reps: 10 - 20 - 30 - 40 - 50 - 60 - 70 - 80 - 90 - 100 - 110 - 120 - 130
140 - 150 - 160 - 170 - 180 - 190 - 200 - 210 - 220 - 230 - 240 - 250 - 260
270 - 280 - 290 - 300 - 310 - 320 - 330 - 340 - 350 - 360 - 370 - 380 - 390
400 - 410 - 420 - 430 - 440 - 450 - 460 - 470 - 480 - 490 - 500 - 510 - 520
530 - 540 - 550 - 560 - 570 - 580 - 590 - 600 - 610 - 620 - 630 - 640 - 650
660 - 670 - 680 - 690 - 700 - 710 - 720 - 730 - 740 - 750 - 760 - 770 - 780
790 - 800 - 810 - 820 - 830 - 840 - 850 - 860 - 870 - 880 - 890 - 900 - 910
920 - 930 - 940 - 950 - 960 - 970 - 980 - 990 - 1000

Notes: ______________________________

____

____

______________________________ **End Date:** ____________

Defense Against:___________________________ **Start Date:** __________

Notes:____

____

____

____

Slow Reps: 10 - 20 - 30 - 40 - 50 - 60 - 70 - 80 - 90 - 100 - 110 - 120 - 130
140 - 150 - 160 - 170 - 180 - 190 - 200 - 210 - 220 - 230 - 240 - 250 - 260
270 - 280 - 290 - 300 - 310 - 320 - 330 - 340 - 350 - 360 - 370 - 380 - 390
400 - 410 - 420 - 430 - 440 - 450 - 460 - 470 - 480 - 490 - 500 - 510 - 520
530 - 540 - 550 - 560 - 570 - 580 - 590 - 600 - 610 - 620 - 630 - 640 - 650
660 - 670 - 680 - 690 - 700 - 710 - 720 - 730 - 740 - 750 - 760 - 770 - 780
790 - 800 - 810 - 820 - 830 - 840 - 850 - 860 - 870 - 880 - 890 - 900 - 910
920 - 930 - 940 - 950 - 960 - 970 - 980 - 990 - 1000

Reg Reps: 10 - 20 - 30 - 40 - 50 - 60 - 70 - 80 - 90 - 100 - 110 - 120 - 130
140 - 150 - 160 - 170 - 180 - 190 - 200 - 210 - 220 - 230 - 240 - 250 - 260
270 - 280 - 290 - 300 - 310 - 320 - 330 - 340 - 350 - 360 - 370 - 380 - 390
400 - 410 - 420 - 430 - 440 - 450 - 460 - 470 - 480 - 490 - 500 - 510 - 520
530 - 540 - 550 - 560 - 570 - 580 - 590 - 600 - 610 - 620 - 630 - 640 - 650
660 - 670 - 680 - 690 - 700 - 710 - 720 - 730 - 740 - 750 - 760 - 770 - 780
790 - 800 - 810 - 820 - 830 - 840 - 850 - 860 - 870 - 880 - 890 - 900 - 910
920 - 930 - 940 - 950 - 960 - 970 - 980 - 990 - 1000

Notes: ____

____

____

____

Reg Reps: 10 - 20 - 30 - 40 - 50 - 60 - 70 - 80 - 90 - 100 - 110 - 120 - 130
140 - 150 - 160 - 170 - 180 - 190 - 200 - 210 - 220 - 230 - 240 - 250 - 260
270 - 280 - 290 - 300 - 310 - 320 - 330 - 340 - 350 - 360 - 370 - 380 - 390
400 - 410 - 420 - 430 - 440 - 450 - 460 - 470 - 480 - 490 - 500 - 510 - 520
530 - 540 - 550 - 560 - 570 - 580 - 590 - 600 - 610 - 620 - 630 - 640 - 650
660 - 670 - 680 - 690 - 700 - 710 - 720 - 730 - 740 - 750 - 760 - 770 - 780
790 - 800 - 810 - 820 - 830 - 840 - 850 - 860 - 870 - 880 - 890 - 900 - 910
920 - 930 - 940 - 950 - 960 - 970 - 980 - 990 - 1000

Notes: ____

____

____

____

Reg Reps: 10 - 20 - 30 - 40 - 50 - 60 - 70 - 80 - 90 - 100 - 110 - 120 - 130
140 - 150 - 160 - 170 - 180 - 190 - 200 - 210 - 220 - 230 - 240 - 250 - 260
270 - 280 - 290 - 300 - 310 - 320 - 330 - 340 - 350 - 360 - 370 - 380 - 390
400 - 410 - 420 - 430 - 440 - 450 - 460 - 470 - 480 - 490 - 500 - 510 - 520
530 - 540 - 550 - 560 - 570 - 580 - 590 - 600 - 610 - 620 - 630 - 640 - 650
660 - 670 - 680 - 690 - 700 - 710 - 720 - 730 - 740 - 750 - 760 - 770 - 780
790 - 800 - 810 - 820 - 830 - 840 - 850 - 860 - 870 - 880 - 890 - 900 - 910
920 - 930 - 940 - 950 - 960 - 970 - 980 - 990 - 1000

Notes: ____________________

Reg Reps: 10 - 20 - 30 - 40 - 50 - 60 - 70 - 80 - 90 - 100 - 110 - 120 - 130
140 - 150 - 160 - 170 - 180 - 190 - 200 - 210 - 220 - 230 - 240 - 250 - 260
270 - 280 - 290 - 300 - 310 - 320 - 330 - 340 - 350 - 360 - 370 - 380 - 390
400 - 410 - 420 - 430 - 440 - 450 - 460 - 470 - 480 - 490 - 500 - 510 - 520
530 - 540 - 550 - 560 - 570 - 580 - 590 - 600 - 610 - 620 - 630 - 640 - 650
660 - 670 - 680 - 690 - 700 - 710 - 720 - 730 - 740 - 750 - 760 - 770 - 780
790 - 800 - 810 - 820 - 830 - 840 - 850 - 860 - 870 - 880 - 890 - 900 - 910
920 - 930 - 940 - 950 - 960 - 970 - 980 - 990 - 1000

Notes: ____________________

Reg Reps: 10 - 20 - 30 - 40 - 50 - 60 - 70 - 80 - 90 - 100 - 110 - 120 - 130
140 - 150 - 160 - 170 - 180 - 190 - 200 - 210 - 220 - 230 - 240 - 250 - 260
270 - 280 - 290 - 300 - 310 - 320 - 330 - 340 - 350 - 360 - 370 - 380 - 390
400 - 410 - 420 - 430 - 440 - 450 - 460 - 470 - 480 - 490 - 500 - 510 - 520
530 - 540 - 550 - 560 - 570 - 580 - 590 - 600 - 610 - 620 - 630 - 640 - 650
660 - 670 - 680 - 690 - 700 - 710 - 720 - 730 - 740 - 750 - 760 - 770 - 780
790 - 800 - 810 - 820 - 830 - 840 - 850 - 860 - 870 - 880 - 890 - 900 - 910
920 - 930 - 940 - 950 - 960 - 970 - 980 - 990 - 1000

Notes: ____________________

End Date: ____________

Defense Against:______________________ **Start Date:** __________

Notes:____

____

____

____

Slow Reps: 10 - 20 - 30 - 40 - 50 - 60 - 70 - 80 - 90 - 100 - 110 - 120 - 130
140 - 150 - 160 - 170 - 180 - 190 - 200 - 210 - 220 - 230 - 240 - 250 - 260
270 - 280 - 290 - 300 - 310 - 320 - 330 - 340 - 350 - 360 - 370 - 380 - 390
400 - 410 - 420 - 430 - 440 - 450 - 460 - 470 - 480 - 490 - 500 - 510 - 520
530 - 540 - 550 - 560 - 570 - 580 - 590 - 600 - 610 - 620 - 630 - 640 - 650
660 - 670 - 680 - 690 - 700 - 710 - 720 - 730 - 740 - 750 - 760 - 770 - 780
790 - 800 - 810 - 820 - 830 - 840 - 850 - 860 - 870 - 880 - 890 - 900 - 910
920 - 930 - 940 - 950 - 960 - 970 - 980 - 990 - 1000

Reg Reps: 10 - 20 - 30 - 40 - 50 - 60 - 70 - 80 - 90 - 100 - 110 - 120 - 130
140 - 150 - 160 - 170 - 180 - 190 - 200 - 210 - 220 - 230 - 240 - 250 - 260
270 - 280 - 290 - 300 - 310 - 320 - 330 - 340 - 350 - 360 - 370 - 380 - 390
400 - 410 - 420 - 430 - 440 - 450 - 460 - 470 - 480 - 490 - 500 - 510 - 520
530 - 540 - 550 - 560 - 570 - 580 - 590 - 600 - 610 - 620 - 630 - 640 - 650
660 - 670 - 680 - 690 - 700 - 710 - 720 - 730 - 740 - 750 - 760 - 770 - 780
790 - 800 - 810 - 820 - 830 - 840 - 850 - 860 - 870 - 880 - 890 - 900 - 910
920 - 930 - 940 - 950 - 960 - 970 - 980 - 990 - 1000

Notes: ____

____

____

____

Reg Reps: 10 - 20 - 30 - 40 - 50 - 60 - 70 - 80 - 90 - 100 - 110 - 120 - 130
140 - 150 - 160 - 170 - 180 - 190 - 200 - 210 - 220 - 230 - 240 - 250 - 260
270 - 280 - 290 - 300 - 310 - 320 - 330 - 340 - 350 - 360 - 370 - 380 - 390
400 - 410 - 420 - 430 - 440 - 450 - 460 - 470 - 480 - 490 - 500 - 510 - 520
530 - 540 - 550 - 560 - 570 - 580 - 590 - 600 - 610 - 620 - 630 - 640 - 650
660 - 670 - 680 - 690 - 700 - 710 - 720 - 730 - 740 - 750 - 760 - 770 - 780
790 - 800 - 810 - 820 - 830 - 840 - 850 - 860 - 870 - 880 - 890 - 900 - 910
920 - 930 - 940 - 950 - 960 - 970 - 980 - 990 - 1000

Notes: ____

____

____

____

Reg Reps: 10 - 20 - 30 - 40 - 50 - 60 - 70 - 80 - 90 - 100 - 110 - 120 - 130
140 - 150 - 160 - 170 - 180 - 190 - 200 - 210 - 220 - 230 - 240 - 250 - 260
270 - 280 - 290 - 300 - 310 - 320 - 330 - 340 - 350 - 360 - 370 - 380 - 390
400 - 410 - 420 - 430 - 440 - 450 - 460 - 470 - 480 - 490 - 500 - 510 - 520
530 - 540 - 550 - 560 - 570 - 580 - 590 - 600 - 610 - 620 - 630 - 640 - 650
660 - 670 - 680 - 690 - 700 - 710 - 720 - 730 - 740 - 750 - 760 - 770 - 780
790 - 800 - 810 - 820 - 830 - 840 - 850 - 860 - 870 - 880 - 890 - 900 - 910
920 - 930 - 940 - 950 - 960 - 970 - 980 - 990 - 1000

Notes: ______________________________

____

____

____

Reg Reps: 10 - 20 - 30 - 40 - 50 - 60 - 70 - 80 - 90 - 100 - 110 - 120 - 130
140 - 150 - 160 - 170 - 180 - 190 - 200 - 210 - 220 - 230 - 240 - 250 - 260
270 - 280 - 290 - 300 - 310 - 320 - 330 - 340 - 350 - 360 - 370 - 380 - 390
400 - 410 - 420 - 430 - 440 - 450 - 460 - 470 - 480 - 490 - 500 - 510 - 520
530 - 540 - 550 - 560 - 570 - 580 - 590 - 600 - 610 - 620 - 630 - 640 - 650
660 - 670 - 680 - 690 - 700 - 710 - 720 - 730 - 740 - 750 - 760 - 770 - 780
790 - 800 - 810 - 820 - 830 - 840 - 850 - 860 - 870 - 880 - 890 - 900 - 910
920 - 930 - 940 - 950 - 960 - 970 - 980 - 990 - 1000

Notes: ______________________________

____

____

____

Reg Reps: 10 - 20 - 30 - 40 - 50 - 60 - 70 - 80 - 90 - 100 - 110 - 120 - 130
140 - 150 - 160 - 170 - 180 - 190 - 200 - 210 - 220 - 230 - 240 - 250 - 260
270 - 280 - 290 - 300 - 310 - 320 - 330 - 340 - 350 - 360 - 370 - 380 - 390
400 - 410 - 420 - 430 - 440 - 450 - 460 - 470 - 480 - 490 - 500 - 510 - 520
530 - 540 - 550 - 560 - 570 - 580 - 590 - 600 - 610 - 620 - 630 - 640 - 650
660 - 670 - 680 - 690 - 700 - 710 - 720 - 730 - 740 - 750 - 760 - 770 - 780
790 - 800 - 810 - 820 - 830 - 840 - 850 - 860 - 870 - 880 - 890 - 900 - 910
920 - 930 - 940 - 950 - 960 - 970 - 980 - 990 - 1000

Notes: ______________________________

____

____

______________________________ **End Date:** ____________

Defense Against:__________________________ **Start Date:** __________

Notes:____

Slow Reps: 10 - 20 - 30 - 40 - 50 - 60 - 70 - 80 - 90 - 100 - 110 - 120 - 130
140 - 150 - 160 - 170 - 180 - 190 - 200 - 210 - 220 - 230 - 240 - 250 - 260
270 - 280 - 290 - 300 - 310 - 320 - 330 - 340 - 350 - 360 - 370 - 380 - 390
400 - 410 - 420 - 430 - 440 - 450 - 460 - 470 - 480 - 490 - 500 - 510 - 520
530 - 540 - 550 - 560 - 570 - 580 - 590 - 600 - 610 - 620 - 630 - 640 - 650
660 - 670 - 680 - 690 - 700 - 710 - 720 - 730 - 740 - 750 - 760 - 770 - 780
790 - 800 - 810 - 820 - 830 - 840 - 850 - 860 - 870 - 880 - 890 - 900 - 910
920 - 930 - 940 - 950 - 960 - 970 - 980 - 990 - 1000

Reg Reps: 10 - 20 - 30 - 40 - 50 - 60 - 70 - 80 - 90 - 100 - 110 - 120 - 130
140 - 150 - 160 - 170 - 180 - 190 - 200 - 210 - 220 - 230 - 240 - 250 - 260
270 - 280 - 290 - 300 - 310 - 320 - 330 - 340 - 350 - 360 - 370 - 380 - 390
400 - 410 - 420 - 430 - 440 - 450 - 460 - 470 - 480 - 490 - 500 - 510 - 520
530 - 540 - 550 - 560 - 570 - 580 - 590 - 600 - 610 - 620 - 630 - 640 - 650
660 - 670 - 680 - 690 - 700 - 710 - 720 - 730 - 740 - 750 - 760 - 770 - 780
790 - 800 - 810 - 820 - 830 - 840 - 850 - 860 - 870 - 880 - 890 - 900 - 910
920 - 930 - 940 - 950 - 960 - 970 - 980 - 990 - 1000

Notes: ____

Reg Reps: 10 - 20 - 30 - 40 - 50 - 60 - 70 - 80 - 90 - 100 - 110 - 120 - 130
140 - 150 - 160 - 170 - 180 - 190 - 200 - 210 - 220 - 230 - 240 - 250 - 260
270 - 280 - 290 - 300 - 310 - 320 - 330 - 340 - 350 - 360 - 370 - 380 - 390
400 - 410 - 420 - 430 - 440 - 450 - 460 - 470 - 480 - 490 - 500 - 510 - 520
530 - 540 - 550 - 560 - 570 - 580 - 590 - 600 - 610 - 620 - 630 - 640 - 650
660 - 670 - 680 - 690 - 700 - 710 - 720 - 730 - 740 - 750 - 760 - 770 - 780
790 - 800 - 810 - 820 - 830 - 840 - 850 - 860 - 870 - 880 - 890 - 900 - 910
920 - 930 - 940 - 950 - 960 - 970 - 980 - 990 - 1000

Notes: ____

Reg Reps: 10 - 20 - 30 - 40 - 50 - 60 - 70 - 80 - 90 - 100 - 110 - 120 - 130
140 - 150 - 160 - 170 - 180 - 190 - 200 - 210 - 220 - 230 - 240 - 250 - 260
270 - 280 - 290 - 300 - 310 - 320 - 330 - 340 - 350 - 360 - 370 - 380 - 390
400 - 410 - 420 - 430 - 440 - 450 - 460 - 470 - 480 - 490 - 500 - 510 - 520
530 - 540 - 550 - 560 - 570 - 580 - 590 - 600 - 610 - 620 - 630 - 640 - 650
660 - 670 - 680 - 690 - 700 - 710 - 720 - 730 - 740 - 750 - 760 - 770 - 780
790 - 800 - 810 - 820 - 830 - 840 - 850 - 860 - 870 - 880 - 890 - 900 - 910
920 - 930 - 940 - 950 - 960 - 970 - 980 - 990 - 1000

Notes: ______________________________

____

____

____

Reg Reps: 10 - 20 - 30 - 40 - 50 - 60 - 70 - 80 - 90 - 100 - 110 - 120 - 130
140 - 150 - 160 - 170 - 180 - 190 - 200 - 210 - 220 - 230 - 240 - 250 - 260
270 - 280 - 290 - 300 - 310 - 320 - 330 - 340 - 350 - 360 - 370 - 380 - 390
400 - 410 - 420 - 430 - 440 - 450 - 460 - 470 - 480 - 490 - 500 - 510 - 520
530 - 540 - 550 - 560 - 570 - 580 - 590 - 600 - 610 - 620 - 630 - 640 - 650
660 - 670 - 680 - 690 - 700 - 710 - 720 - 730 - 740 - 750 - 760 - 770 - 780
790 - 800 - 810 - 820 - 830 - 840 - 850 - 860 - 870 - 880 - 890 - 900 - 910
920 - 930 - 940 - 950 - 960 - 970 - 980 - 990 - 1000

Notes: ______________________________

____

____

____

Reg Reps: 10 - 20 - 30 - 40 - 50 - 60 - 70 - 80 - 90 - 100 - 110 - 120 - 130
140 - 150 - 160 - 170 - 180 - 190 - 200 - 210 - 220 - 230 - 240 - 250 - 260
270 - 280 - 290 - 300 - 310 - 320 - 330 - 340 - 350 - 360 - 370 - 380 - 390
400 - 410 - 420 - 430 - 440 - 450 - 460 - 470 - 480 - 490 - 500 - 510 - 520
530 - 540 - 550 - 560 - 570 - 580 - 590 - 600 - 610 - 620 - 630 - 640 - 650
660 - 670 - 680 - 690 - 700 - 710 - 720 - 730 - 740 - 750 - 760 - 770 - 780
790 - 800 - 810 - 820 - 830 - 840 - 850 - 860 - 870 - 880 - 890 - 900 - 910
920 - 930 - 940 - 950 - 960 - 970 - 980 - 990 - 1000

Notes: ______________________________

____

____

______________________________ **End Date:** ____________

Defense Against:________________________ **Start Date:** __________
Notes:____

Slow Reps: 10 - 20 - 30 - 40 - 50 - 60 - 70 - 80 - 90 - 100 - 110 - 120 - 130
140 - 150 - 160 - 170 - 180 - 190 - 200 - 210 - 220 - 230 - 240 - 250 - 260
270 - 280 - 290 - 300 - 310 - 320 - 330 - 340 - 350 - 360 - 370 - 380 - 390
400 - 410 - 420 - 430 - 440 - 450 - 460 - 470 - 480 - 490 - 500 - 510 - 520
530 - 540 - 550 - 560 - 570 - 580 - 590 - 600 - 610 - 620 - 630 - 640 - 650
660 - 670 - 680 - 690 - 700 - 710 - 720 - 730 - 740 - 750 - 760 - 770 - 780
790 - 800 - 810 - 820 - 830 - 840 - 850 - 860 - 870 - 880 - 890 - 900 - 910
920 - 930 - 940 - 950 - 960 - 970 - 980 - 990 - 1000

Reg Reps: 10 - 20 - 30 - 40 - 50 - 60 - 70 - 80 - 90 - 100 - 110 - 120 - 130
140 - 150 - 160 - 170 - 180 - 190 - 200 - 210 - 220 - 230 - 240 - 250 - 260
270 - 280 - 290 - 300 - 310 - 320 - 330 - 340 - 350 - 360 - 370 - 380 - 390
400 - 410 - 420 - 430 - 440 - 450 - 460 - 470 - 480 - 490 - 500 - 510 - 520
530 - 540 - 550 - 560 - 570 - 580 - 590 - 600 - 610 - 620 - 630 - 640 - 650
660 - 670 - 680 - 690 - 700 - 710 - 720 - 730 - 740 - 750 - 760 - 770 - 780
790 - 800 - 810 - 820 - 830 - 840 - 850 - 860 - 870 - 880 - 890 - 900 - 910
920 - 930 - 940 - 950 - 960 - 970 - 980 - 990 - 1000
Notes: ____

Reg Reps: 10 - 20 - 30 - 40 - 50 - 60 - 70 - 80 - 90 - 100 - 110 - 120 - 130
140 - 150 - 160 - 170 - 180 - 190 - 200 - 210 - 220 - 230 - 240 - 250 - 260
270 - 280 - 290 - 300 - 310 - 320 - 330 - 340 - 350 - 360 - 370 - 380 - 390
400 - 410 - 420 - 430 - 440 - 450 - 460 - 470 - 480 - 490 - 500 - 510 - 520
530 - 540 - 550 - 560 - 570 - 580 - 590 - 600 - 610 - 620 - 630 - 640 - 650
660 - 670 - 680 - 690 - 700 - 710 - 720 - 730 - 740 - 750 - 760 - 770 - 780
790 - 800 - 810 - 820 - 830 - 840 - 850 - 860 - 870 - 880 - 890 - 900 - 910
920 - 930 - 940 - 950 - 960 - 970 - 980 - 990 - 1000
Notes: ____

Reg Reps: 10 - 20 - 30 - 40 - 50 - 60 - 70 - 80 - 90 - 100 - 110 - 120 - 130
140 - 150 - 160 - 170 - 180 - 190 - 200 - 210 - 220 - 230 - 240 - 250 - 260
270 - 280 - 290 - 300 - 310 - 320 - 330 - 340 - 350 - 360 - 370 - 380 - 390
400 - 410 - 420 - 430 - 440 - 450 - 460 - 470 - 480 - 490 - 500 - 510 - 520
530 - 540 - 550 - 560 - 570 - 580 - 590 - 600 - 610 - 620 - 630 - 640 - 650
660 - 670 - 680 - 690 - 700 - 710 - 720 - 730 - 740 - 750 - 760 - 770 - 780
790 - 800 - 810 - 820 - 830 - 840 - 850 - 860 - 870 - 880 - 890 - 900 - 910
920 - 930 - 940 - 950 - 960 - 970 - 980 - 990 - 1000

Notes: ____________________

Reg Reps: 10 - 20 - 30 - 40 - 50 - 60 - 70 - 80 - 90 - 100 - 110 - 120 - 130
140 - 150 - 160 - 170 - 180 - 190 - 200 - 210 - 220 - 230 - 240 - 250 - 260
270 - 280 - 290 - 300 - 310 - 320 - 330 - 340 - 350 - 360 - 370 - 380 - 390
400 - 410 - 420 - 430 - 440 - 450 - 460 - 470 - 480 - 490 - 500 - 510 - 520
530 - 540 - 550 - 560 - 570 - 580 - 590 - 600 - 610 - 620 - 630 - 640 - 650
660 - 670 - 680 - 690 - 700 - 710 - 720 - 730 - 740 - 750 - 760 - 770 - 780
790 - 800 - 810 - 820 - 830 - 840 - 850 - 860 - 870 - 880 - 890 - 900 - 910
920 - 930 - 940 - 950 - 960 - 970 - 980 - 990 - 1000

Notes: ____________________

Reg Reps: 10 - 20 - 30 - 40 - 50 - 60 - 70 - 80 - 90 - 100 - 110 - 120 - 130
140 - 150 - 160 - 170 - 180 - 190 - 200 - 210 - 220 - 230 - 240 - 250 - 260
270 - 280 - 290 - 300 - 310 - 320 - 330 - 340 - 350 - 360 - 370 - 380 - 390
400 - 410 - 420 - 430 - 440 - 450 - 460 - 470 - 480 - 490 - 500 - 510 - 520
530 - 540 - 550 - 560 - 570 - 580 - 590 - 600 - 610 - 620 - 630 - 640 - 650
660 - 670 - 680 - 690 - 700 - 710 - 720 - 730 - 740 - 750 - 760 - 770 - 780
790 - 800 - 810 - 820 - 830 - 840 - 850 - 860 - 870 - 880 - 890 - 900 - 910
920 - 930 - 940 - 950 - 960 - 970 - 980 - 990 - 1000

Notes: ____________________

End Date: ____________

Defense Against: ______________________ **Start Date:** __________

Notes: ____

____

____

____

Slow Reps: 10 - 20 - 30 - 40 - 50 - 60 - 70 - 80 - 90 - 100 - 110 - 120 - 130
140 - 150 - 160 - 170 - 180 - 190 - 200 - 210 - 220 - 230 - 240 - 250 - 260
270 - 280 - 290 - 300 - 310 - 320 - 330 - 340 - 350 - 360 - 370 - 380 - 390
400 - 410 - 420 - 430 - 440 - 450 - 460 - 470 - 480 - 490 - 500 - 510 - 520
530 - 540 - 550 - 560 - 570 - 580 - 590 - 600 - 610 - 620 - 630 - 640 - 650
660 - 670 - 680 - 690 - 700 - 710 - 720 - 730 - 740 - 750 - 760 - 770 - 780
790 - 800 - 810 - 820 - 830 - 840 - 850 - 860 - 870 - 880 - 890 - 900 - 910
920 - 930 - 940 - 950 - 960 - 970 - 980 - 990 - 1000

Reg Reps: 10 - 20 - 30 - 40 - 50 - 60 - 70 - 80 - 90 - 100 - 110 - 120 - 130
140 - 150 - 160 - 170 - 180 - 190 - 200 - 210 - 220 - 230 - 240 - 250 - 260
270 - 280 - 290 - 300 - 310 - 320 - 330 - 340 - 350 - 360 - 370 - 380 - 390
400 - 410 - 420 - 430 - 440 - 450 - 460 - 470 - 480 - 490 - 500 - 510 - 520
530 - 540 - 550 - 560 - 570 - 580 - 590 - 600 - 610 - 620 - 630 - 640 - 650
660 - 670 - 680 - 690 - 700 - 710 - 720 - 730 - 740 - 750 - 760 - 770 - 780
790 - 800 - 810 - 820 - 830 - 840 - 850 - 860 - 870 - 880 - 890 - 900 - 910
920 - 930 - 940 - 950 - 960 - 970 - 980 - 990 - 1000

Notes: ____

____

____

____

Reg Reps: 10 - 20 - 30 - 40 - 50 - 60 - 70 - 80 - 90 - 100 - 110 - 120 - 130
140 - 150 - 160 - 170 - 180 - 190 - 200 - 210 - 220 - 230 - 240 - 250 - 260
270 - 280 - 290 - 300 - 310 - 320 - 330 - 340 - 350 - 360 - 370 - 380 - 390
400 - 410 - 420 - 430 - 440 - 450 - 460 - 470 - 480 - 490 - 500 - 510 - 520
530 - 540 - 550 - 560 - 570 - 580 - 590 - 600 - 610 - 620 - 630 - 640 - 650
660 - 670 - 680 - 690 - 700 - 710 - 720 - 730 - 740 - 750 - 760 - 770 - 780
790 - 800 - 810 - 820 - 830 - 840 - 850 - 860 - 870 - 880 - 890 - 900 - 910
920 - 930 - 940 - 950 - 960 - 970 - 980 - 990 - 1000

Notes: ____

____

____

____

Reg Reps: 10 - 20 - 30 - 40 - 50 - 60 - 70 - 80 - 90 - 100 - 110 - 120 - 130
140 - 150 - 160 - 170 - 180 - 190 - 200 - 210 - 220 - 230 - 240 - 250 - 260
270 - 280 - 290 - 300 - 310 - 320 - 330 - 340 - 350 - 360 - 370 - 380 - 390
400 - 410 - 420 - 430 - 440 - 450 - 460 - 470 - 480 - 490 - 500 - 510 - 520
530 - 540 - 550 - 560 - 570 - 580 - 590 - 600 - 610 - 620 - 630 - 640 - 650
660 - 670 - 680 - 690 - 700 - 710 - 720 - 730 - 740 - 750 - 760 - 770 - 780
790 - 800 - 810 - 820 - 830 - 840 - 850 - 860 - 870 - 880 - 890 - 900 - 910
920 - 930 - 940 - 950 - 960 - 970 - 980 - 990 - 1000

Notes: ______________________________

____

____

____

Reg Reps: 10 - 20 - 30 - 40 - 50 - 60 - 70 - 80 - 90 - 100 - 110 - 120 - 130
140 - 150 - 160 - 170 - 180 - 190 - 200 - 210 - 220 - 230 - 240 - 250 - 260
270 - 280 - 290 - 300 - 310 - 320 - 330 - 340 - 350 - 360 - 370 - 380 - 390
400 - 410 - 420 - 430 - 440 - 450 - 460 - 470 - 480 - 490 - 500 - 510 - 520
530 - 540 - 550 - 560 - 570 - 580 - 590 - 600 - 610 - 620 - 630 - 640 - 650
660 - 670 - 680 - 690 - 700 - 710 - 720 - 730 - 740 - 750 - 760 - 770 - 780
790 - 800 - 810 - 820 - 830 - 840 - 850 - 860 - 870 - 880 - 890 - 900 - 910
920 - 930 - 940 - 950 - 960 - 970 - 980 - 990 - 1000

Notes: ______________________________

____

____

____

Reg Reps: 10 - 20 - 30 - 40 - 50 - 60 - 70 - 80 - 90 - 100 - 110 - 120 - 130
140 - 150 - 160 - 170 - 180 - 190 - 200 - 210 - 220 - 230 - 240 - 250 - 260
270 - 280 - 290 - 300 - 310 - 320 - 330 - 340 - 350 - 360 - 370 - 380 - 390
400 - 410 - 420 - 430 - 440 - 450 - 460 - 470 - 480 - 490 - 500 - 510 - 520
530 - 540 - 550 - 560 - 570 - 580 - 590 - 600 - 610 - 620 - 630 - 640 - 650
660 - 670 - 680 - 690 - 700 - 710 - 720 - 730 - 740 - 750 - 760 - 770 - 780
790 - 800 - 810 - 820 - 830 - 840 - 850 - 860 - 870 - 880 - 890 - 900 - 910
920 - 930 - 940 - 950 - 960 - 970 - 980 - 990 - 1000

Notes: ______________________________

____

____

______________________________ **End Date:** __________

Defense Against:__________________________ **Start Date:** __________

Notes:____

Slow Reps: 10 - 20 - 30 - 40 - 50 - 60 - 70 - 80 - 90 - 100 - 110 - 120 - 130
140 - 150 - 160 - 170 - 180 - 190 - 200 - 210 - 220 - 230 - 240 - 250 - 260
270 - 280 - 290 - 300 - 310 - 320 - 330 - 340 - 350 - 360 - 370 - 380 - 390
400 - 410 - 420 - 430 - 440 - 450 - 460 - 470 - 480 - 490 - 500 - 510 - 520
530 - 540 - 550 - 560 - 570 - 580 - 590 - 600 - 610 - 620 - 630 - 640 - 650
660 - 670 - 680 - 690 - 700 - 710 - 720 - 730 - 740 - 750 - 760 - 770 - 780
790 - 800 - 810 - 820 - 830 - 840 - 850 - 860 - 870 - 880 - 890 - 900 - 910
920 - 930 - 940 - 950 - 960 - 970 - 980 - 990 - 1000

Reg Reps: 10 - 20 - 30 - 40 - 50 - 60 - 70 - 80 - 90 - 100 - 110 - 120 - 130
140 - 150 - 160 - 170 - 180 - 190 - 200 - 210 - 220 - 230 - 240 - 250 - 260
270 - 280 - 290 - 300 - 310 - 320 - 330 - 340 - 350 - 360 - 370 - 380 - 390
400 - 410 - 420 - 430 - 440 - 450 - 460 - 470 - 480 - 490 - 500 - 510 - 520
530 - 540 - 550 - 560 - 570 - 580 - 590 - 600 - 610 - 620 - 630 - 640 - 650
660 - 670 - 680 - 690 - 700 - 710 - 720 - 730 - 740 - 750 - 760 - 770 - 780
790 - 800 - 810 - 820 - 830 - 840 - 850 - 860 - 870 - 880 - 890 - 900 - 910
920 - 930 - 940 - 950 - 960 - 970 - 980 - 990 - 1000

Notes: ____

Reg Reps: 10 - 20 - 30 - 40 - 50 - 60 - 70 - 80 - 90 - 100 - 110 - 120 - 130
140 - 150 - 160 - 170 - 180 - 190 - 200 - 210 - 220 - 230 - 240 - 250 - 260
270 - 280 - 290 - 300 - 310 - 320 - 330 - 340 - 350 - 360 - 370 - 380 - 390
400 - 410 - 420 - 430 - 440 - 450 - 460 - 470 - 480 - 490 - 500 - 510 - 520
530 - 540 - 550 - 560 - 570 - 580 - 590 - 600 - 610 - 620 - 630 - 640 - 650
660 - 670 - 680 - 690 - 700 - 710 - 720 - 730 - 740 - 750 - 760 - 770 - 780
790 - 800 - 810 - 820 - 830 - 840 - 850 - 860 - 870 - 880 - 890 - 900 - 910
920 - 930 - 940 - 950 - 960 - 970 - 980 - 990 - 1000

Notes: ____

Reg Reps: 10 - 20 - 30 - 40 - 50 - 60 - 70 - 80 - 90 - 100 - 110 - 120 - 130
140 - 150 - 160 - 170 - 180 - 190 - 200 - 210 - 220 - 230 - 240 - 250 - 260
270 - 280 - 290 - 300 - 310 - 320 - 330 - 340 - 350 - 360 - 370 - 380 - 390
400 - 410 - 420 - 430 - 440 - 450 - 460 - 470 - 480 - 490 - 500 - 510 - 520
530 - 540 - 550 - 560 - 570 - 580 - 590 - 600 - 610 - 620 - 630 - 640 - 650
660 - 670 - 680 - 690 - 700 - 710 - 720 - 730 - 740 - 750 - 760 - 770 - 780
790 - 800 - 810 - 820 - 830 - 840 - 850 - 860 - 870 - 880 - 890 - 900 - 910
920 - 930 - 940 - 950 - 960 - 970 - 980 - 990 - 1000

Notes: ____

____

____

____

Reg Reps: 10 - 20 - 30 - 40 - 50 - 60 - 70 - 80 - 90 - 100 - 110 - 120 - 130
140 - 150 - 160 - 170 - 180 - 190 - 200 - 210 - 220 - 230 - 240 - 250 - 260
270 - 280 - 290 - 300 - 310 - 320 - 330 - 340 - 350 - 360 - 370 - 380 - 390
400 - 410 - 420 - 430 - 440 - 450 - 460 - 470 - 480 - 490 - 500 - 510 - 520
530 - 540 - 550 - 560 - 570 - 580 - 590 - 600 - 610 - 620 - 630 - 640 - 650
660 - 670 - 680 - 690 - 700 - 710 - 720 - 730 - 740 - 750 - 760 - 770 - 780
790 - 800 - 810 - 820 - 830 - 840 - 850 - 860 - 870 - 880 - 890 - 900 - 910
920 - 930 - 940 - 950 - 960 - 970 - 980 - 990 - 1000

Notes: ____

____

____

____

Reg Reps: 10 - 20 - 30 - 40 - 50 - 60 - 70 - 80 - 90 - 100 - 110 - 120 - 130
140 - 150 - 160 - 170 - 180 - 190 - 200 - 210 - 220 - 230 - 240 - 250 - 260
270 - 280 - 290 - 300 - 310 - 320 - 330 - 340 - 350 - 360 - 370 - 380 - 390
400 - 410 - 420 - 430 - 440 - 450 - 460 - 470 - 480 - 490 - 500 - 510 - 520
530 - 540 - 550 - 560 - 570 - 580 - 590 - 600 - 610 - 620 - 630 - 640 - 650
660 - 670 - 680 - 690 - 700 - 710 - 720 - 730 - 740 - 750 - 760 - 770 - 780
790 - 800 - 810 - 820 - 830 - 840 - 850 - 860 - 870 - 880 - 890 - 900 - 910
920 - 930 - 940 - 950 - 960 - 970 - 980 - 990 - 1000

Notes: ____

____

____

______________________________ **End Date:** ____________

Defense Against: ______________________ **Start Date:** __________

Notes: ____

____

____

____

Slow Reps: 10 - 20 - 30 - 40 - 50 - 60 - 70 - 80 - 90 - 100 - 110 - 120 - 130
140 - 150 - 160 - 170 - 180 - 190 - 200 - 210 - 220 - 230 - 240 - 250 - 260
270 - 280 - 290 - 300 - 310 - 320 - 330 - 340 - 350 - 360 - 370 - 380 - 390
400 - 410 - 420 - 430 - 440 - 450 - 460 - 470 - 480 - 490 - 500 - 510 - 520
530 - 540 - 550 - 560 - 570 - 580 - 590 - 600 - 610 - 620 - 630 - 640 - 650
660 - 670 - 680 - 690 - 700 - 710 - 720 - 730 - 740 - 750 - 760 - 770 - 780
790 - 800 - 810 - 820 - 830 - 840 - 850 - 860 - 870 - 880 - 890 - 900 - 910
920 - 930 - 940 - 950 - 960 - 970 - 980 - 990 - 1000

Reg Reps: 10 - 20 - 30 - 40 - 50 - 60 - 70 - 80 - 90 - 100 - 110 - 120 - 130
140 - 150 - 160 - 170 - 180 - 190 - 200 - 210 - 220 - 230 - 240 - 250 - 260
270 - 280 - 290 - 300 - 310 - 320 - 330 - 340 - 350 - 360 - 370 - 380 - 390
400 - 410 - 420 - 430 - 440 - 450 - 460 - 470 - 480 - 490 - 500 - 510 - 520
530 - 540 - 550 - 560 - 570 - 580 - 590 - 600 - 610 - 620 - 630 - 640 - 650
660 - 670 - 680 - 690 - 700 - 710 - 720 - 730 - 740 - 750 - 760 - 770 - 780
790 - 800 - 810 - 820 - 830 - 840 - 850 - 860 - 870 - 880 - 890 - 900 - 910
920 - 930 - 940 - 950 - 960 - 970 - 980 - 990 - 1000

Notes: ____

____

____

____

Reg Reps: 10 - 20 - 30 - 40 - 50 - 60 - 70 - 80 - 90 - 100 - 110 - 120 - 130
140 - 150 - 160 - 170 - 180 - 190 - 200 - 210 - 220 - 230 - 240 - 250 - 260
270 - 280 - 290 - 300 - 310 - 320 - 330 - 340 - 350 - 360 - 370 - 380 - 390
400 - 410 - 420 - 430 - 440 - 450 - 460 - 470 - 480 - 490 - 500 - 510 - 520
530 - 540 - 550 - 560 - 570 - 580 - 590 - 600 - 610 - 620 - 630 - 640 - 650
660 - 670 - 680 - 690 - 700 - 710 - 720 - 730 - 740 - 750 - 760 - 770 - 780
790 - 800 - 810 - 820 - 830 - 840 - 850 - 860 - 870 - 880 - 890 - 900 - 910
920 - 930 - 940 - 950 - 960 - 970 - 980 - 990 - 1000

Notes: ____

____

____

____

Reg Reps: 10 - 20 - 30 - 40 - 50 - 60 - 70 - 80 - 90 - 100 - 110 - 120 - 130
140 - 150 - 160 - 170 - 180 - 190 - 200 - 210 - 220 - 230 - 240 - 250 - 260
270 - 280 - 290 - 300 - 310 - 320 - 330 - 340 - 350 - 360 - 370 - 380 - 390
400 - 410 - 420 - 430 - 440 - 450 - 460 - 470 - 480 - 490 - 500 - 510 - 520
530 - 540 - 550 - 560 - 570 - 580 - 590 - 600 - 610 - 620 - 630 - 640 - 650
660 - 670 - 680 - 690 - 700 - 710 - 720 - 730 - 740 - 750 - 760 - 770 - 780
790 - 800 - 810 - 820 - 830 - 840 - 850 - 860 - 870 - 880 - 890 - 900 - 910
920 - 930 - 940 - 950 - 960 - 970 - 980 - 990 - 1000

Notes: ____

____

____

____

Reg Reps: 10 - 20 - 30 - 40 - 50 - 60 - 70 - 80 - 90 - 100 - 110 - 120 - 130
140 - 150 - 160 - 170 - 180 - 190 - 200 - 210 - 220 - 230 - 240 - 250 - 260
270 - 280 - 290 - 300 - 310 - 320 - 330 - 340 - 350 - 360 - 370 - 380 - 390
400 - 410 - 420 - 430 - 440 - 450 - 460 - 470 - 480 - 490 - 500 - 510 - 520
530 - 540 - 550 - 560 - 570 - 580 - 590 - 600 - 610 - 620 - 630 - 640 - 650
660 - 670 - 680 - 690 - 700 - 710 - 720 - 730 - 740 - 750 - 760 - 770 - 780
790 - 800 - 810 - 820 - 830 - 840 - 850 - 860 - 870 - 880 - 890 - 900 - 910
920 - 930 - 940 - 950 - 960 - 970 - 980 - 990 - 1000

Notes: ____

____

____

____

Reg Reps: 10 - 20 - 30 - 40 - 50 - 60 - 70 - 80 - 90 - 100 - 110 - 120 - 130
140 - 150 - 160 - 170 - 180 - 190 - 200 - 210 - 220 - 230 - 240 - 250 - 260
270 - 280 - 290 - 300 - 310 - 320 - 330 - 340 - 350 - 360 - 370 - 380 - 390
400 - 410 - 420 - 430 - 440 - 450 - 460 - 470 - 480 - 490 - 500 - 510 - 520
530 - 540 - 550 - 560 - 570 - 580 - 590 - 600 - 610 - 620 - 630 - 640 - 650
660 - 670 - 680 - 690 - 700 - 710 - 720 - 730 - 740 - 750 - 760 - 770 - 780
790 - 800 - 810 - 820 - 830 - 840 - 850 - 860 - 870 - 880 - 890 - 900 - 910
920 - 930 - 940 - 950 - 960 - 970 - 980 - 990 - 1000

Notes: ____

____

____

______________________________ **End Date:** ____________

Defense Against:____________________________ **Start Date:** ___________

Notes:____

Slow Reps: 10 - 20 - 30 - 40 - 50 - 60 - 70 - 80 - 90 - 100 - 110 - 120 - 130
140 - 150 - 160 - 170 - 180 - 190 - 200 - 210 - 220 - 230 - 240 - 250 - 260
270 - 280 - 290 - 300 - 310 - 320 - 330 - 340 - 350 - 360 - 370 - 380 - 390
400 - 410 - 420 - 430 - 440 - 450 - 460 - 470 - 480 - 490 - 500 - 510 - 520
530 - 540 - 550 - 560 - 570 - 580 - 590 - 600 - 610 - 620 - 630 - 640 - 650
660 - 670 - 680 - 690 - 700 - 710 - 720 - 730 - 740 - 750 - 760 - 770 - 780
790 - 800 - 810 - 820 - 830 - 840 - 850 - 860 - 870 - 880 - 890 - 900 - 910
920 - 930 - 940 - 950 - 960 - 970 - 980 - 990 - 1000

Reg Reps: 10 - 20 - 30 - 40 - 50 - 60 - 70 - 80 - 90 - 100 - 110 - 120 - 130
140 - 150 - 160 - 170 - 180 - 190 - 200 - 210 - 220 - 230 - 240 - 250 - 260
270 - 280 - 290 - 300 - 310 - 320 - 330 - 340 - 350 - 360 - 370 - 380 - 390
400 - 410 - 420 - 430 - 440 - 450 - 460 - 470 - 480 - 490 - 500 - 510 - 520
530 - 540 - 550 - 560 - 570 - 580 - 590 - 600 - 610 - 620 - 630 - 640 - 650
660 - 670 - 680 - 690 - 700 - 710 - 720 - 730 - 740 - 750 - 760 - 770 - 780
790 - 800 - 810 - 820 - 830 - 840 - 850 - 860 - 870 - 880 - 890 - 900 - 910
920 - 930 - 940 - 950 - 960 - 970 - 980 - 990 - 1000

Notes: ____

Reg Reps: 10 - 20 - 30 - 40 - 50 - 60 - 70 - 80 - 90 - 100 - 110 - 120 - 130
140 - 150 - 160 - 170 - 180 - 190 - 200 - 210 - 220 - 230 - 240 - 250 - 260
270 - 280 - 290 - 300 - 310 - 320 - 330 - 340 - 350 - 360 - 370 - 380 - 390
400 - 410 - 420 - 430 - 440 - 450 - 460 - 470 - 480 - 490 - 500 - 510 - 520
530 - 540 - 550 - 560 - 570 - 580 - 590 - 600 - 610 - 620 - 630 - 640 - 650
660 - 670 - 680 - 690 - 700 - 710 - 720 - 730 - 740 - 750 - 760 - 770 - 780
790 - 800 - 810 - 820 - 830 - 840 - 850 - 860 - 870 - 880 - 890 - 900 - 910
920 - 930 - 940 - 950 - 960 - 970 - 980 - 990 - 1000

Notes: ____

Reg Reps: 10 - 20 - 30 - 40 - 50 - 60 - 70 - 80 - 90 - 100 - 110 - 120 - 130
140 - 150 - 160 - 170 - 180 - 190 - 200 - 210 - 220 - 230 - 240 - 250 - 260
270 - 280 - 290 - 300 - 310 - 320 - 330 - 340 - 350 - 360 - 370 - 380 - 390
400 - 410 - 420 - 430 - 440 - 450 - 460 - 470 - 480 - 490 - 500 - 510 - 520
530 - 540 - 550 - 560 - 570 - 580 - 590 - 600 - 610 - 620 - 630 - 640 - 650
660 - 670 - 680 - 690 - 700 - 710 - 720 - 730 - 740 - 750 - 760 - 770 - 780
790 - 800 - 810 - 820 - 830 - 840 - 850 - 860 - 870 - 880 - 890 - 900 - 910
920 - 930 - 940 - 950 - 960 - 970 - 980 - 990 - 1000

Notes: ____________________

Reg Reps: 10 - 20 - 30 - 40 - 50 - 60 - 70 - 80 - 90 - 100 - 110 - 120 - 130
140 - 150 - 160 - 170 - 180 - 190 - 200 - 210 - 220 - 230 - 240 - 250 - 260
270 - 280 - 290 - 300 - 310 - 320 - 330 - 340 - 350 - 360 - 370 - 380 - 390
400 - 410 - 420 - 430 - 440 - 450 - 460 - 470 - 480 - 490 - 500 - 510 - 520
530 - 540 - 550 - 560 - 570 - 580 - 590 - 600 - 610 - 620 - 630 - 640 - 650
660 - 670 - 680 - 690 - 700 - 710 - 720 - 730 - 740 - 750 - 760 - 770 - 780
790 - 800 - 810 - 820 - 830 - 840 - 850 - 860 - 870 - 880 - 890 - 900 - 910
920 - 930 - 940 - 950 - 960 - 970 - 980 - 990 - 1000

Notes: ____________________

Reg Reps: 10 - 20 - 30 - 40 - 50 - 60 - 70 - 80 - 90 - 100 - 110 - 120 - 130
140 - 150 - 160 - 170 - 180 - 190 - 200 - 210 - 220 - 230 - 240 - 250 - 260
270 - 280 - 290 - 300 - 310 - 320 - 330 - 340 - 350 - 360 - 370 - 380 - 390
400 - 410 - 420 - 430 - 440 - 450 - 460 - 470 - 480 - 490 - 500 - 510 - 520
530 - 540 - 550 - 560 - 570 - 580 - 590 - 600 - 610 - 620 - 630 - 640 - 650
660 - 670 - 680 - 690 - 700 - 710 - 720 - 730 - 740 - 750 - 760 - 770 - 780
790 - 800 - 810 - 820 - 830 - 840 - 850 - 860 - 870 - 880 - 890 - 900 - 910
920 - 930 - 940 - 950 - 960 - 970 - 980 - 990 - 1000

Notes: ____________________

End Date: __________

Defense Against:__________________________ **Start Date:** __________

Notes:____

Slow Reps: 10 - 20 - 30 - 40 - 50 - 60 - 70 - 80 - 90 - 100 - 110 - 120 - 130
140 - 150 - 160 - 170 - 180 - 190 - 200 - 210 - 220 - 230 - 240 - 250 - 260
270 - 280 - 290 - 300 - 310 - 320 - 330 - 340 - 350 - 360 - 370 - 380 - 390
400 - 410 - 420 - 430 - 440 - 450 - 460 - 470 - 480 - 490 - 500 - 510 - 520
530 - 540 - 550 - 560 - 570 - 580 - 590 - 600 - 610 - 620 - 630 - 640 - 650
660 - 670 - 680 - 690 - 700 - 710 - 720 - 730 - 740 - 750 - 760 - 770 - 780
790 - 800 - 810 - 820 - 830 - 840 - 850 - 860 - 870 - 880 - 890 - 900 - 910
920 - 930 - 940 - 950 - 960 - 970 - 980 - 990 - 1000

Reg Reps: 10 - 20 - 30 - 40 - 50 - 60 - 70 - 80 - 90 - 100 - 110 - 120 - 130
140 - 150 - 160 - 170 - 180 - 190 - 200 - 210 - 220 - 230 - 240 - 250 - 260
270 - 280 - 290 - 300 - 310 - 320 - 330 - 340 - 350 - 360 - 370 - 380 - 390
400 - 410 - 420 - 430 - 440 - 450 - 460 - 470 - 480 - 490 - 500 - 510 - 520
530 - 540 - 550 - 560 - 570 - 580 - 590 - 600 - 610 - 620 - 630 - 640 - 650
660 - 670 - 680 - 690 - 700 - 710 - 720 - 730 - 740 - 750 - 760 - 770 - 780
790 - 800 - 810 - 820 - 830 - 840 - 850 - 860 - 870 - 880 - 890 - 900 - 910
920 - 930 - 940 - 950 - 960 - 970 - 980 - 990 - 1000

Notes: ____

Reg Reps: 10 - 20 - 30 - 40 - 50 - 60 - 70 - 80 - 90 - 100 - 110 - 120 - 130
140 - 150 - 160 - 170 - 180 - 190 - 200 - 210 - 220 - 230 - 240 - 250 - 260
270 - 280 - 290 - 300 - 310 - 320 - 330 - 340 - 350 - 360 - 370 - 380 - 390
400 - 410 - 420 - 430 - 440 - 450 - 460 - 470 - 480 - 490 - 500 - 510 - 520
530 - 540 - 550 - 560 - 570 - 580 - 590 - 600 - 610 - 620 - 630 - 640 - 650
660 - 670 - 680 - 690 - 700 - 710 - 720 - 730 - 740 - 750 - 760 - 770 - 780
790 - 800 - 810 - 820 - 830 - 840 - 850 - 860 - 870 - 880 - 890 - 900 - 910
920 - 930 - 940 - 950 - 960 - 970 - 980 - 990 - 1000

Notes: ____

Reg Reps: 10 - 20 - 30 - 40 - 50 - 60 - 70 - 80 - 90 - 100 - 110 - 120 - 130
140 - 150 - 160 - 170 - 180 - 190 - 200 - 210 - 220 - 230 - 240 - 250 - 260
270 - 280 - 290 - 300 - 310 - 320 - 330 - 340 - 350 - 360 - 370 - 380 - 390
400 - 410 - 420 - 430 - 440 - 450 - 460 - 470 - 480 - 490 - 500 - 510 - 520
530 - 540 - 550 - 560 - 570 - 580 - 590 - 600 - 610 - 620 - 630 - 640 - 650
660 - 670 - 680 - 690 - 700 - 710 - 720 - 730 - 740 - 750 - 760 - 770 - 780
790 - 800 - 810 - 820 - 830 - 840 - 850 - 860 - 870 - 880 - 890 - 900 - 910
920 - 930 - 940 - 950 - 960 - 970 - 980 - 990 - 1000

Notes: ______________________________

____

____

____

Reg Reps: 10 - 20 - 30 - 40 - 50 - 60 - 70 - 80 - 90 - 100 - 110 - 120 - 130
140 - 150 - 160 - 170 - 180 - 190 - 200 - 210 - 220 - 230 - 240 - 250 - 260
270 - 280 - 290 - 300 - 310 - 320 - 330 - 340 - 350 - 360 - 370 - 380 - 390
400 - 410 - 420 - 430 - 440 - 450 - 460 - 470 - 480 - 490 - 500 - 510 - 520
530 - 540 - 550 - 560 - 570 - 580 - 590 - 600 - 610 - 620 - 630 - 640 - 650
660 - 670 - 680 - 690 - 700 - 710 - 720 - 730 - 740 - 750 - 760 - 770 - 780
790 - 800 - 810 - 820 - 830 - 840 - 850 - 860 - 870 - 880 - 890 - 900 - 910
920 - 930 - 940 - 950 - 960 - 970 - 980 - 990 - 1000

Notes: ______________________________

____

____

____

Reg Reps: 10 - 20 - 30 - 40 - 50 - 60 - 70 - 80 - 90 - 100 - 110 - 120 - 130
140 - 150 - 160 - 170 - 180 - 190 - 200 - 210 - 220 - 230 - 240 - 250 - 260
270 - 280 - 290 - 300 - 310 - 320 - 330 - 340 - 350 - 360 - 370 - 380 - 390
400 - 410 - 420 - 430 - 440 - 450 - 460 - 470 - 480 - 490 - 500 - 510 - 520
530 - 540 - 550 - 560 - 570 - 580 - 590 - 600 - 610 - 620 - 630 - 640 - 650
660 - 670 - 680 - 690 - 700 - 710 - 720 - 730 - 740 - 750 - 760 - 770 - 780
790 - 800 - 810 - 820 - 830 - 840 - 850 - 860 - 870 - 880 - 890 - 900 - 910
920 - 930 - 940 - 950 - 960 - 970 - 980 - 990 - 1000

Notes: ______________________________

____

____

______________________________ **End Date:** ______________

Defense Against: ______________________ **Start Date:** __________
Notes: ____

____

____

____

Slow Reps: 10 - 20 - 30 - 40 - 50 - 60 - 70 - 80 - 90 - 100 - 110 - 120 - 130
140 - 150 - 160 - 170 - 180 - 190 - 200 - 210 - 220 - 230 - 240 - 250 - 260
270 - 280 - 290 - 300 - 310 - 320 - 330 - 340 - 350 - 360 - 370 - 380 - 390
400 - 410 - 420 - 430 - 440 - 450 - 460 - 470 - 480 - 490 - 500 - 510 - 520
530 - 540 - 550 - 560 - 570 - 580 - 590 - 600 - 610 - 620 - 630 - 640 - 650
660 - 670 - 680 - 690 - 700 - 710 - 720 - 730 - 740 - 750 - 760 - 770 - 780
790 - 800 - 810 - 820 - 830 - 840 - 850 - 860 - 870 - 880 - 890 - 900 - 910
920 - 930 - 940 - 950 - 960 - 970 - 980 - 990 - 1000

Reg Reps: 10 - 20 - 30 - 40 - 50 - 60 - 70 - 80 - 90 - 100 - 110 - 120 - 130
140 - 150 - 160 - 170 - 180 - 190 - 200 - 210 - 220 - 230 - 240 - 250 - 260
270 - 280 - 290 - 300 - 310 - 320 - 330 - 340 - 350 - 360 - 370 - 380 - 390
400 - 410 - 420 - 430 - 440 - 450 - 460 - 470 - 480 - 490 - 500 - 510 - 520
530 - 540 - 550 - 560 - 570 - 580 - 590 - 600 - 610 - 620 - 630 - 640 - 650
660 - 670 - 680 - 690 - 700 - 710 - 720 - 730 - 740 - 750 - 760 - 770 - 780
790 - 800 - 810 - 820 - 830 - 840 - 850 - 860 - 870 - 880 - 890 - 900 - 910
920 - 930 - 940 - 950 - 960 - 970 - 980 - 990 - 1000
Notes: ____

____

____

____

Reg Reps: 10 - 20 - 30 - 40 - 50 - 60 - 70 - 80 - 90 - 100 - 110 - 120 - 130
140 - 150 - 160 - 170 - 180 - 190 - 200 - 210 - 220 - 230 - 240 - 250 - 260
270 - 280 - 290 - 300 - 310 - 320 - 330 - 340 - 350 - 360 - 370 - 380 - 390
400 - 410 - 420 - 430 - 440 - 450 - 460 - 470 - 480 - 490 - 500 - 510 - 520
530 - 540 - 550 - 560 - 570 - 580 - 590 - 600 - 610 - 620 - 630 - 640 - 650
660 - 670 - 680 - 690 - 700 - 710 - 720 - 730 - 740 - 750 - 760 - 770 - 780
790 - 800 - 810 - 820 - 830 - 840 - 850 - 860 - 870 - 880 - 890 - 900 - 910
920 - 930 - 940 - 950 - 960 - 970 - 980 - 990 - 1000
Notes: ____

____

____

____

Reg Reps: 10 - 20 - 30 - 40 - 50 - 60 - 70 - 80 - 90 - 100 - 110 - 120 - 130 - 140 - 150 - 160 - 170 - 180 - 190 - 200 - 210 - 220 - 230 - 240 - 250 - 260 - 270 - 280 - 290 - 300 - 310 - 320 - 330 - 340 - 350 - 360 - 370 - 380 - 390 - 400 - 410 - 420 - 430 - 440 - 450 - 460 - 470 - 480 - 490 - 500 - 510 - 520 - 530 - 540 - 550 - 560 - 570 - 580 - 590 - 600 - 610 - 620 - 630 - 640 - 650 - 660 - 670 - 680 - 690 - 700 - 710 - 720 - 730 - 740 - 750 - 760 - 770 - 780 - 790 - 800 - 810 - 820 - 830 - 840 - 850 - 860 - 870 - 880 - 890 - 900 - 910 - 920 - 930 - 940 - 950 - 960 - 970 - 980 - 990 - 1000

Notes: ______________________________

____

____

____

Reg Reps: 10 - 20 - 30 - 40 - 50 - 60 - 70 - 80 - 90 - 100 - 110 - 120 - 130 - 140 - 150 - 160 - 170 - 180 - 190 - 200 - 210 - 220 - 230 - 240 - 250 - 260 - 270 - 280 - 290 - 300 - 310 - 320 - 330 - 340 - 350 - 360 - 370 - 380 - 390 - 400 - 410 - 420 - 430 - 440 - 450 - 460 - 470 - 480 - 490 - 500 - 510 - 520 - 530 - 540 - 550 - 560 - 570 - 580 - 590 - 600 - 610 - 620 - 630 - 640 - 650 - 660 - 670 - 680 - 690 - 700 - 710 - 720 - 730 - 740 - 750 - 760 - 770 - 780 - 790 - 800 - 810 - 820 - 830 - 840 - 850 - 860 - 870 - 880 - 890 - 900 - 910 - 920 - 930 - 940 - 950 - 960 - 970 - 980 - 990 - 1000

Notes: ______________________________

____

____

____

Reg Reps: 10 - 20 - 30 - 40 - 50 - 60 - 70 - 80 - 90 - 100 - 110 - 120 - 130 - 140 - 150 - 160 - 170 - 180 - 190 - 200 - 210 - 220 - 230 - 240 - 250 - 260 - 270 - 280 - 290 - 300 - 310 - 320 - 330 - 340 - 350 - 360 - 370 - 380 - 390 - 400 - 410 - 420 - 430 - 440 - 450 - 460 - 470 - 480 - 490 - 500 - 510 - 520 - 530 - 540 - 550 - 560 - 570 - 580 - 590 - 600 - 610 - 620 - 630 - 640 - 650 - 660 - 670 - 680 - 690 - 700 - 710 - 720 - 730 - 740 - 750 - 760 - 770 - 780 - 790 - 800 - 810 - 820 - 830 - 840 - 850 - 860 - 870 - 880 - 890 - 900 - 910 - 920 - 930 - 940 - 950 - 960 - 970 - 980 - 990 - 1000

Notes: ______________________________

____

____

______________________________ **End Date:** __________

ADDITIONAL NOTES

Previous Page:_______ Defense Against:______________________ Rep #:_____

Previous Page:_______ Defense Against:______________________ Rep #:_____

Previous Page:_______ Defense Against:______________________ Rep #:_____

Previous Page:_______ Defense Against:______________________ Rep #:_____

ADDITIONAL NOTES

Previous Page:_______ Defense Against:______________________ Rep #:_____

Previous Page:_______ Defense Against:______________________ Rep #:_____

Previous Page:_______ Defense Against:______________________ Rep #:_____

Previous Page:_______ Defense Against:______________________ Rep #:_____

ADDITIONAL NOTES

Previous Page:_______ Defense Against:_________________________ Rep #:_____

Previous Page:_______ Defense Against:_________________________ Rep #:_____

Previous Page:_______ Defense Against:_________________________ Rep #:_____

Previous Page:_______ Defense Against:_________________________ Rep #:_____

ADDITIONAL NOTES

Previous Page:_______ Defense Against:_______________________ Rep #:_____

Previous Page:_______ Defense Against:_______________________ Rep #:_____

Previous Page:_______ Defense Against:_______________________ Rep #:_____

Previous Page:_______ Defense Against:_______________________ Rep #:_____

ADDITIONAL NOTES

Previous Page:_______ Defense Against:_______________________ Rep #:_____

Previous Page:_______ Defense Against:_______________________ Rep #:_____

Previous Page:_______ Defense Against:_______________________ Rep #:_____

Previous Page:_______ Defense Against:_______________________ Rep #:_____

ADDITIONAL NOTES

Previous Page:_______ Defense Against:_______________________ Rep #:_____

Previous Page:_______ Defense Against:_______________________ Rep #:_____

Previous Page:_______ Defense Against:_______________________ Rep #:_____

Previous Page:_______ Defense Against:_______________________ Rep #:_____

ADDITIONAL NOTES

Previous Page:_______ Defense Against:_________________________ Rep #:_____

Previous Page:_______ Defense Against:_________________________ Rep #:_____

Previous Page:_______ Defense Against:_________________________ Rep #:_____

Previous Page:_______ Defense Against:_________________________ Rep #:_____

ADDITIONAL NOTES

Previous Page:_______ Defense Against:_______________________ Rep #:_____

Previous Page:_______ Defense Against:_______________________ Rep #:_____

Previous Page:_______ Defense Against:_______________________ Rep #:_____

Previous Page:_______ Defense Against:_______________________ Rep #:_____

ADDITIONAL NOTES

Previous Page:_______ Defense Against:______________________ Rep #:_____

Previous Page:_______ Defense Against:______________________ Rep #:_____

Previous Page:_______ Defense Against:______________________ Rep #:_____

Previous Page:_______ Defense Against:______________________ Rep #:_____

ADDITIONAL NOTES

Previous Page:_______ Defense Against:________________________ Rep #:_____

Previous Page:_______ Defense Against:________________________ Rep #:_____

Previous Page:_______ Defense Against:________________________ Rep #:_____

Previous Page:_______ Defense Against:________________________ Rep #:_____

ADDITIONAL NOTES

Previous Page:_______ Defense Against:_________________________ Rep #:_____

Previous Page:_______ Defense Against:_________________________ Rep #:_____

Previous Page:_______ Defense Against:_________________________ Rep #:_____

Previous Page:_______ Defense Against:_________________________ Rep #:_____

ADDITIONAL NOTES

Previous Page:______ Defense Against:____________________ Rep #:_____

Previous Page:______ Defense Against:____________________ Rep #:_____

Previous Page:______ Defense Against:____________________ Rep #:_____

Previous Page:______ Defense Against:____________________ Rep #:_____

ADDITIONAL NOTES

Previous Page:_______ Defense Against:_______________________ Rep #:_____

Previous Page:_______ Defense Against:_______________________ Rep #:_____

Previous Page:_______ Defense Against:_______________________ Rep #:_____

Previous Page:_______ Defense Against:_______________________ Rep #:_____

ADDITIONAL NOTES

Previous Page:_______ Defense Against:_______________________ Rep #:_____

Previous Page:_______ Defense Against:_______________________ Rep #:_____

Previous Page:_______ Defense Against:_______________________ Rep #:_____

Previous Page:_______ Defense Against:_______________________ Rep #:_____

ADDITIONAL NOTES

Previous Page:_______ Defense Against:_________________________ Rep #:_____

Previous Page:_______ Defense Against:_________________________ Rep #:_____

Previous Page:_______ Defense Against:_________________________ Rep #:_____

Previous Page:_______ Defense Against:_________________________ Rep #:_____

ADDITIONAL NOTES

Previous Page:_______ Defense Against:______________________ Rep #:_____

Previous Page:_______ Defense Against:______________________ Rep #:_____

Previous Page:_______ Defense Against:______________________ Rep #:_____

Previous Page:_______ Defense Against:______________________ Rep #:_____

Made in United States
Troutdale, OR
11/03/2026

56844425R00073